The Data Science Framework

Juan J. Cuadrado-Gallego • Yuri Demchenko
Editors

The Data Science Framework

A View from the EDISON Project

 Springer

Editors
Juan J. Cuadrado-Gallego
Department of Computer Science
University of Alcalá
Madrid, Spain

Yuri Demchenko
Faculty of Science
Universiteit van Amsterdam
Amsterdam, The Netherlands

ISBN 978-3-030-51025-1   ISBN 978-3-030-51023-7   (eBook)
https://doi.org/10.1007/978-3-030-51023-7

This Springer imprint is published by the registered company Springer Nature Switzerland AG.
The registered company address is: Gewerbestrasse 11, 6330 Cham, Switzerland

Preface

From Juan J. Cuadrado-Gallego and Yuri Demchenko

We want to give our acknowledgements to the European Commission that under the code 675419 funds the EDISON project between the years 2015 and 2017. We acknowledge all project members of the EDISON project and community involved who created facilitating and motivating environment for the EDSF development and implementation. Also, we would also like to acknowledge our universities, Universiteit van Amsterdam, Amsterdam, the Netherlands, and the Universidad de Alcalá, Madrid, Spain, for their support during the realisation of this book. And finally we acknowledge the rest of researchers who have collaborated in the realisation of any of the chapters of the book: Luca Comminiello, from the Universidad degli Studi de Perugia, Italy; Tomasz Wiktorski, from the Universitetet i Stavanger, Norway; Oleg Chertov, from the Igor Sikorsky Kyiv Polytechnic Institute of the National Technical University of Ukraine, Ukraine; Ernestina Menasalvas, Ana M. Moreno and Nik Swoboda, from the Universidad Politécnica de Madrid, Spain; and Steve Brewer, from the University of Southampton, UK.

From Juan J. Cuadrado-Gallego

This is the first important work that I have been able to do after having suffered a major accident in 2016 from which it was really very difficult to recover. I want to thank and dedicate this book to all the people without whom it would have been impossible to get it, especially to my wife and my daughters, Ana Pérez de la Serna, Ana Cuadrado and Cristina Cuadrado.

From Yuri Demchenko

I dedicate this book to my lovely women, my wife Natalia, my daughter Anastasia and my granddaughter Sonia. All of them have an important role in what I do, what I write, what I learn and how I live. While my wife helped me to understand some elements of business data management, my granddaughter is helping me to understand how to teach data science starting from the school.

Amsterdam, The Netherlands

Juan J. Cuadrado-Gallego
Yuri Demchenko

January, 2020

v

Contents

Acronyms

ACM	Association for Computer Machinery
BA	Business analytics
BI	Business intelligence
BOK	Body of knowledge
BPM	Business process management
BABOK	Business analysis body of knowledge
CCS	Classification Computer Science by ACM
CCECC	Committee for Computing Education in Community Colleges
CEN	European Committee for Standardization
CF-DS	Data Science Competence Framework
CODATA	International Council for Science: Committee on Data for Science and Technology
CRISP-DM	Cross-Industry Standard Process for Data Mining
CS	Computer science
CWA	CEN Workshop Agreement
DigComp	Digital competences for citizens (EU report 2017)
DM-BoK	Data Management Body of Knowledge by DAMAI
DS-BoK	Data Science Body of Knowledge
DSCP	Data Science Community Portal
DSPP	Data Science Professional Profiles
DSPS	Data Science Professional Skills
EC	European Commission
e-CF	European e-Competence Framework
EDSA	European Data Science Academy
EDSF	EDISON Data Science Framework
EGI	European Grid Initiative
ELG	EDISON Liaison Group
EOEE	EDISON Online E-Learning Environment
EOSC	European Open Science Cloud
EQF	European Qualification Framework

ERA	European Research Area
ESCO	European Skills, Competences, Qualifications and Occupations
ETM-DS	Data Science Education and Training Model
EU	European Union
EUA	European Association for Data Science
EUDAT	http://eudat.eu/what-eudat
HPCS	High Performance Computing and Simulation Conference
ICT	Information and Communications Technologies
IEEE	Institute of Electrical and Electronics Engineers
IG-ETHRD	RDA Interest Group on Education and Training on Handling Research Data
IPR	Intellectual property rights
ISCO	International Standard Classification of Occupations
KA	Knowledge area
KAG	Knowledge area group
KU	Knowledge unit
LERU	League of European Research Universities
LIBER	Association of European Research Libraries
LO	Learning outcome
LU	Learning unit
MC-DS	Data Science Model Curriculum
NIST	National Institute of Standards and Technology of the US Department of Commerce
NBD-WG	NIST Big Data Working Group
NBDIF	NIST Big Data Interoperability Framework
P21	21st-Century Skills Framework
PID	Persistent identifier
PM-BoK	Project Management Body of Knowledge
PRACE	Partnership for Advanced Computing in Europe
RDA	Research Data Alliance
SWEBOK	Software Engineering Body of Knowledge
UK	United Kingdom
US	United States

List of Figures

Chapter 1
Introduction to the Data Science Framework

Juan J. Cuadrado-Gallego and Yuri Demchenko

This initial chapter, "Introduction to a data science framework", presents the main concepts related to the subject of the book. Starting with the common word *about*, the chapter presents an introduction to data science, the main subject of the book; to EDISON, the European Union- (EU) funded project under which most of the contents of this book were developed; and to the data science framework proposal developed in the EDISON project, and a more in-depth explanation, in the following chapters, will be the larger part of the book. The chapter will end with a last about, in this case about this book, in which the contents and structure of the book will be briefly presented.

1.1   About Data Science

What Is Data Science?
All the contents of this book have been thought with the goal to give to the reader the most solid answer to that question, from many points of views: its theoretical foundations, its practical use, its professional practice and others.

There are multiple definitions of the data science discipline and technology that stress/put in the centre one of the four flavours/goal of data analysis:

J. J. Cuadrado-Gallego (✉)
Department of Computer Science, University of Alcalá, Madrid, Spain
e-mail: jjcg@uah.es

Y. Demchenko (✉)
Universiteit van Amsterdam, Amsterdam, The Netherlands
e-mail: y.demchenko@uva.nl

© Springer Nature Switzerland AG 2020

J. J. Cuadrado-Gallego, Y. Demchenko (eds.), *The Data Science Framework*,
https://doi.org/10.1007/978-3-030-51023-7_1

- *Data analytics* is a process of inspecting, transforming and modelling data with the goal to discover trends, patterns or relations that describe observable real-life phenomena and can be used for informed decision-making.
- *Data science* makes the systematic study of the structure and behaviour of data in order to understand past and current occurrences, as well as predict the future behaviour of that data. Data science is an interdisciplinary field that uses scientific methods, processes, algorithms and systems to extract knowledge and insights from structured and unstructured data.
- *Machine learning* deals with the development of algorithms, some of them based on statistical models, with the objective that their computational implementation allows the computer not only to carry out the tasks without supervision but learn of the results for a continuous improvement. Within machine learning, *deep learning* is the set of predictive methodologies that use artificial neural networks to progressively extract higher level features from unstructured raw data. This class of methods is particularly effective for making predictions from big amounts of data generated by real-life behavioural processes or sensors. Machine learning and deep learning are considered as subfields of the data science focused on specific tasks, while data science provides general methodology for working with a wide variety of data using different methods and tools.
- *Artificial intelligence* is a machine or application with the capability to autonomously execute upon predictions it makes from data, where prediction is made based on data science and analytics methods.

It is important to clarify the relation of the data science to other closely related scientific disciplines and technology domains such as *big data, artificial intelligence, machine learning* and *statistics*. Despite the fact that some authors may refer to historical facts of mentioning these many years ago [1], we refer to the current data-driven technologies development that made data science a central component of all other data-related and data-driven technologies development. We identify that such technology fusion and consolidation took place in 2011–2013 with the advent of cloud computing and big data which also aligned with the US National Institute of Standards and Technologies' (NIST) [2] definition of cloud computing in 2011 and big data definition in 2013 [3].

Big data serves as a technology platform to allow the data science and analytics solutions and applications to work with modern data which are of the *big data 3V scale*: *volume*, amount of data processed; *velocity*, speed of growing of data processed; and *variety,* number of different types of data processed. Big data technology platform includes large-scale computation, storage and network facilities, typically cloud based, such as Hadoop, Spark, NoSQL databases, data lakes and others.

In the whole digital economy ecosystem, the data science integrates all multiple components from other scientific and technology domains to drive data-intensive research and emerging digital technologies development.

All the contents of this book have been thought to try that, after the reading, the reader can have his or her own answer to the question stated at the beginning of this section, but from the experience of the EDISON Data Science Framework development and with the purpose to have a brief/actionable definition to answer the question, the authors can give the following answer to the question:

What Is Data Science?
Data Science is a complex discipline that uses conceptual and mathematical abstractions and models, statistical methods, together with modern computational tools to obtain knowledge/derive insight from data to (uncover correlations and causations in business data) support decision making in scientific research and business activity.
    Yuri Demchenko and Juan J. Cuadrado-Gallego
    And, if we must define data science in only one sentence:
    Science that studies how to obtain knowledge from Data.
    Juan J. Cuadrado-Gallego and Yuri Demchenko

1.2 About the EDISON Project

The EDISON Project was the EU-funded Horizon 2020 project with the Grant 675419, which was developed since 2015 until 2017, and its goal was to create the foundation for the data science profession for Europe. The EDISON project was originated from the community initiative started at the Research Data Alliance (RDA) [4], with the creation of the RDA Interest Group on Education and Training on Handling Research Data [5] (IG-ETHRD) in 2014, and joined experts and practitioners in research data management to address demand for data specialist that would be capable of bringing value from data explosion at that time. From its start, in September 2015, the project became involved into European Digital Skills Initiative which addressed the whole complex of activities in addressing strong demand for digital and data skills in Europe.

During its term, the EDISON project undertook multiple initiatives and organised multiple community activities/events and conducted important studies to involve data experts and practitioners from academia, research and industry to define the foundation of the new profession of the data scientist.

The main outcome of the EDISON project was, until the publication of this book, the EDISON Data Science Framework (EDSF), which was a product of wide professional community facilitated by the EDISON project. The project published EDSF Release 2 as its final deliverable in 2017. Since the project's end, the EDSF is maintained by the EDISON EDSF Community Initiative [6], coordinated by the University of Amsterdam, and which involves former project partners and numerous contributors from academia, research and industry. The EDSF Release 3 was published in 2018, and new EDSF Release 4 has been published in 2020. This book includes all the knowledge developed for these 4 releases. As the EDSF became mature, the future updates and revisions will be done biannually.

Besides the multidimensional definition of the data science profession, the EDSF created a comprehensive and effective methodology that can be used for other professional domains to address multiple aspects of the organisational human resources management and capacity building that include competences and skills definition and assessment, education and training, customised curriculum design, knowledge assessment and certification, individual professional development and career path building.

1.3 About EDISON Data Science Framework (EDSF)

The EDISON Data Science Framework provides the basis for the definition of the data science profession and enables the definition of the other components related to data science education, training, organisational roles definition and skills management, as well as professional certification.

Figure 1.1 illustrates the main components of the EDISON Data Science Framework and their inter-relations that provide the conceptual basis for the development of the data science profession:

- Data Science Competence Framework (CF-DS). EDSF Part 1.
- Data Science Body of Knowledge (DS-BoK). EDSF Part 2.
- Data Science Model Curriculum (MC-DS). EDSF Part 3.
- Data Science Professional Profiles and Occupations taxonomy (DSPP). EDSF Part 4.
- Data Science Taxonomy and Scientific Disciplines Classification

The proposed framework provides the basis for other components of the data science professional ecosystem (defined and piloted in the EDISON project and constituting the project legacy that can be reused and followed by the community) such as:

- EDISON Online Education Environment (EOEE)
- Education and Training Directory and Marketplace

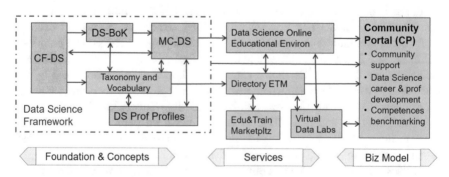

Fig. 1.1 EDISON Data Science Framework components

- Data Science Community Portal (DSCP), which also includes tools for individual competences benchmarking and personalised educational path building
- Certification framework for core Data Science Competences and Professional Profiles.

The *Competences Framework for Data Science* (*CF-DS*) provides the overall basis for the whole EDSF. The core CF-DS includes common competences required for successful work of data scientist in different work environments in industry and in research and through the whole career path. The future CF-DS development will include coverage of the domain-specific competences and skills and will involve domain and subject matter experts.

The *Data Science Body of Knowledge* (*DS-BoK*) defines the knowledge areas for building data science curricula that are required to support identified data science competences. DS-BoK is organised by knowledge area groups (KAG) that correspond to the CF-DS competence groups. Each KAG is composed of knowledge areas (KA). Each KA is composed of a number of knowledge units (KU), which are currently the lowest component of the DS-BoK. DS-BoK incorporates best practices in computer science and domain-specific bodies of knowledge and includes KAs and KUs defined, where possible, based on the classification of computer science components taken from other bodies of knowledge, and proposes new KAs/KUs to incorporate new technologies used in data science and their recent developments.

The *Model Curriculum for Data Science* (*MC-DS*) is built based on CF-DS and DS-BoK where learning outcomes (LO) are defined based on CF-DS competences and learning units (LU) are mapped to knowledge units in DS-BoK. Three mastery, or proficiency, levels are defined for each learning outcome to allow for flexible curricula development and profiling for different Data Science Professional Profiles. The proposed learning outcomes are enumerated to have direct mapping to the enumerated competences in CF-DS.

The *Data Science Professional Profiles* (*DSPP*) are defined as an extension to European Skills, Competences, Occupations and Qualifications (ESCO) [7] taxonomy using the ESCO top classification groups. DSPP definition provides an important instrument to define effective organisational structures and roles related to data science positions and can be also used for building individual career path and corresponding competences and skills transferability between organisations and sectors.

The Data Science Taxonomy and Scientific Disciplines Classification ensures consistency between four core components of EDSF: CF-DS, DS-BoK, MC-DS and DSPP. To ensure consistency and linking between EDSF components, all individual elements of the framework are enumerated, in particular: competences, skills and knowledge topics in CF-DS; knowledge groups, areas and units in DS-BoK; learning outcomes and learning units in MC-DS; and professional profiles in DSPP.

The EDISON data science professional ecosystem illustrated in Fig. 1.1 uses core EDSF components to specify the potential services that can be offered for the professional data science community and provide the basis for the sustainable data science competences and skills management by organisations, in particular in conditions of emerging Industry 4.0, growing digitalisations and artificial intelligence development. As an example of practical use, CF-DS and DS-BoK can be used for individual competences and knowledge benchmarking and play an instrumental role in constructing personalised learning paths and professional (up/re-) skilling programmes based on MC-DS.

The recent EDSF Release 4, treated in this book, which can be referred also as EDSF2020, is the result of cooperation and contribution by the wide community of academicians, researchers and practitioners that are practically involved into data science and data analytics education and training, competences and skills management in organisations and standardisation in the area of competences, skills, occupations and digital technologies.

The EDSF provides the conceptual basis for the data science profession definition, targeted education and training, professional certification, organisational capacity building and organisation and individual skills management and career transferability.

The EDSF Part 5 document, part of the EDSF2020 Release, defines the EDSF use cases and applications:

- Digital competences and data literacy training
- Data science competences analysis and curriculum design
- Assessment of individual and team competences, as well as balanced data science team composition
- Developing tailored curriculum for academic education or professional training to bridge skills gap and staff up/re-skilling

The EDSF Part 5 is intended to provide guidance and the basis for universities, training organisations, data management and data steward team and practitioners to define their data science curricula and courses selection, on one hand, and for companies to better define a set of required competences and skills for their specific industry domain in their search for data science talents, on the other hand.

1.4   About This Book

This book has been written with a double purpose: the first is to gather all the information and knowledge obtained during the development of the EDISON project in a single document that allows a much easier handling by researchers, practitioners, teachers and all those interested in the data science; the second has been to go a step beyond what was obtained in the project and present the knowledge in a more elaborate and expanded way, which allows an easier and deeper assimilation of them. Consequently, the topics are presented in more depth than the reader can find in the documentation resulting from the project, with a substantially revised structure and with a large number of additional documentation. We are firmly sure that all this will be a great improvement for all those interested in data science and who want to know a consolidated framework of it. We hope you enjoy the reading.

Contents are presented in a sequential way in which each chapter takes as starting point the knowledge presented in the previous one. After this initial chapter of introduction, Chap. 2 presents the set of competences that a data scientist must have. Starting from these lists of competences the body of knowledge that the discipline must have to allow us to obtain them is presented in Chap. 3. This body of knowledge is used to define, in Chap. 4, an approach to the development of data science curricula; after the treatment of the knowledge and education of data science seen in the previous chapters, the development of the profession of data scientist is exposed in Chap. 5, in which the Data Science Professional Profiles are presented. Chapter 6 presents a set of four real successful use cases and applications of the EDISON Data Science Framework that can be very useful for the reader in his or her application of the knowledges acquired in the book. The book ends with an Annex in which some models of process related to data science are presented.

Chapter 2
Data Science Competences

Yuri Demchenko and Juan J. Cuadrado-Gallego

This chapter presents the definition of the Data Science Competence Framework (CF-DS). CF-DS is a cornerstone in the definition of the whole data science framework and so it was developed in the EDISON project. CF-DS provides the basis for the Data Science Body of Knowledge (DS-BoK) and Model Curriculum definitions (DS-MC) and further for the Data Science Professional Profiles definition and certification (DSPP). The proposed CF-DS incorporates many of the underpinning principles of the European Union e-Competence Framework (e-CF3.0) [8] and its further standardisation as EN16234-1:2018 [9, 10]. The CF-DS and DSPP have also adopted and intend to comply with the structure of European information and communication technologies (ICT) [11] framework on professional profiles and European Skills, Competences, Occupations and Qualifications (ESCO) [7]. Corresponding information is provided in both documents CF-DS and DSPP.

The presented Data Science Competence Framework definition is based on the analysis of existing frameworks for data science and ICT competences and skills and supported by the analysis of the demand side for data scientist profession in industry and research. The CF-DS has been widely discussed at the numerous workshops, conferences and meetings with wide community contribution. The core CF-DS competences have been reviewed by experts and validated in numerous practical implementations. The following CF-DS development and community contribution was resulted in revising existing and adding new competences, skills and knowledge topics related to new technologies and job market demand for data scientists. The CF-DS also contains the definition of the data science workplace skills that includes

Y. Demchenko
Universiteit van Amsterdam, Amsterdam, The Netherlands
e-mail: y.demchenko@uva.nl

J. J. Cuadrado-Gallego (✉)
Department of Computer Science, University of Alcalá, Madrid, Spain
e-mail: jjcg@uah.es

© Springer Nature Switzerland AG 2020
J. J. Cuadrado-Gallego, Y. Demchenko (eds.), *The Data Science Framework*,
https://doi.org/10.1007/978-3-030-51023-7_2

the data science professional skills, *acting and thinking like data scientist* and the definition of the general *soft skills* often referred to as twenty-first-century skills that are increasingly demanded by modern data-driven companies.

The CF-DS defines five groups of competences for data science that include:

- *Data analytics*
- *Data engineering*
- *Domain knowledge*
- *Data management and governance*
- *Research methods and project management* for research-related occupations, or *business process management* for business-related occupations.

   The research methods competences are essential for the data scientist to discover new relations and provide actionable insight into available data, and have the ability to formulate good research questions and hypothesis and evaluate them based on collected data.

This CF-DS provides the individual competences mapping to identified skills and knowledge topics for each of the five competence groups. The identified competences, skills and knowledge subjects are provided as enumerated lists to allow easy use in applications and developing compatible APIs.

The proposed CF-DS in particular, with the DS-BoK, is intended to provide guidance and the basis for universities to define their data science curricula and courses selection, on one hand, and for companies to better define a set of required competences and skills for their specific industry domain in their search for data science talents, on the other hand.

It is also intended that the proposed CF-DS can provide the basis for building interactive/web-based tool for individual or organisational data science competences benchmarking, data science team building and creating the customised data science education and training programme. The proposed CF-DS can be used for building interactive/web-based tool or applications for knowledge and skills (self-) assessment, job vacancy design and assessment of the candidate's profile for a specific profile/role or job vacancy. All individual competences, knowledge topics and skills are enumerated to allow easier design of API for applications that may use CF-DS. Practical examples of using CF-DS are described in Chap. 6.

This second chapter has three sections with the following structure: Sect. 2.1 provides an overview of existing frameworks for ICT and data science competences and skills definition including NIST 1500-1 Special Publication, e-CF3.0, and ACM Computing Classification System, CCS2012. Section 2.2 presents the full CF-DS definition that includes identified competence groups and identified skills and knowledge that all together should enable the data scientist to effectively work with a variety of data analytics methods and big data platforms to deliver insight and value to organisations. Section 2.3 provides description of the data science common practical skills related to using computational and data management platforms, programming languages and tools, the data science professional workplace and general attitude skills demanded from the modern specialists/professional intended to work in modern agile data-driven companies.

2.1   Overview of Frameworks for Computer Science and Data Science Competences Definitions

This section provides a brief overview of existing standard and commonly accepted frameworks that have been used for defining data science and general computer science and information and communications technologies (ICT) competences, skills and subject domain classifications that can be, with some alignment, built upon and reused for better acceptance from research and industrial communities.

2.1.1   Big Data Interoperability Framework, NIST NBDIF

The NIST Big Data Working Group (NBD-WG) of the US National Institute of Standards and Technologies (NIST) of the US Government Department of Commerce published their first release of NIST Big Data Interoperability Framework (US NBDIF) in September 2015 [3] consisting of 7 volumes. Volume 1. *Definitions* provides some definitions, which are:

- *Data science* is the extraction of actionable knowledge directly from data through a process of discovery, or hypothesis formulation and hypothesis testing. Data science can be understood as the activities happening in the processing layer of the system architecture, against data stored in the data layer, in order to extract knowledge from the raw data.

  Data science across the entire data life cycle incorporates principles, techniques and methods from many disciplines and domains including data cleaning, data management, analytics, visualisation and engineering, and in the context of big data, it now also includes big data engineering. Data science applications implement data transformation processes from the data life cycle in the context of big data engineering.
- *Data scientist* is a practitioner who has enough knowledge in the overlapping regimes of business needs, domain knowledge, analytical skills and software and systems engineering to manage the end-to-end data processes in the data life cycle.

  Data scientists and data science teams solve complex data problems by employing deep expertise in one or more of these disciplines, in the context of business strategy, and under the guidance of domain knowledge. Personal skills in communication, presentation and inquisitiveness are also very important given the complexity of interactions within big data systems.
- The *data life cycle* is the set of processes in an application that transform raw data into actionable knowledge.
- The term *analytics* refers to the discovery of meaningful patterns in data and is one of the steps in the data life cycle of collection of raw data, preparation of information, analysis of patterns to synthesise knowledge and action to produce

Fig. 2.1 Data Science
definition by NIST BDIF [3]

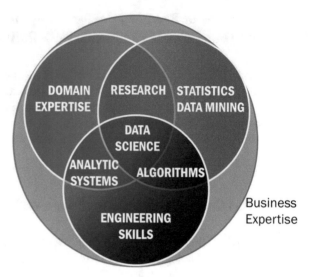

value. Analytics is used to refer to the methods, their implementations in tools and
the results of the use of the tools as interpreted by the practitioner. The analytics
process is the synthesis of knowledge from information.

Figure 2.1 from the US NBDIF publication provides graphical presentation of the
multi-factor/multi-domain data science definition.

The NBDIF Volume 1 also provides overview of other definitions of big data and
data science from IDG, McKinsey, O'Reilly reports and popular blogs published by
experts in a new technology.

NBDIF provided the general approach to the data science competences and skills
definition, in particular, as having three groups: data analytics, data science engineering and domain expertise, which may define possible specialisation of actual
data science curricula or individual data scientists competences profile.

2.1.2   e-Competence Framework, e-CF

The European Commission (EC) of the EU published the e-Competence Framework
(EU e-CF) [8–10], which presents the following approach to the competence
concept:

- Competence is a durable concept, and although technology, jobs, marketing
  terminology and promotional concepts within the ICT environment change
  rapidly, the e-CF remains durable requiring maintenance approximately every
  3 years to maintain relevance.

- A competence can be a part of a job definition but cannot be used to substitute similar named job definition; one single competence can be assigned to multiple job definitions.

Once the definition of competence has been established, the e-CF was also established as a tool to support mutual understanding and provide transparency of language through the articulation of competences required and deployed by ICT professionals, including both practitioners and managers. The e-CF is structured from four dimensions:

- Dimension 1: 5 e-Competence areas, derived from the ICT business processes stages from organisational perspective:

  - *Plan*. Defines activities related to planning services or infrastructure, may include also elements of design and trends monitoring
  - *Build*. Includes activities related to applications development, deployment, engineering and monitoring
  - *Run*. Includes activities to run/operate applications or infrastructure, including user support, change support and problems management
  - *Enable*. Includes numerous activities related to support production and business processes in organisations that include sales support, channel management, knowledge management, personnel development and education and training
  - *Manage*. Includes activities related to ICT/projects and business processes management including management of risk, customer relations and information security

- Dimension 2: A set of reference e-competences for each area, with a generic description for each competence. 40 competences identified in total provide the European generic reference definitions of the e-CF 3.0
- Dimension 3: Proficiency levels of each e-competence provide European reference level specifications on e-competence levels e-1 to e-5, which are related to the European Qualifications Framework (EQF) levels 3 to 8 [12]. Dimension 3 which defines e-competence levels related to the European Qualifications Frameworks (EQF) is a bridge between organisational and individual competences. Refer Chap. 5 that provides mapping between e-CF proficiency levels and EQF qualification levels.
- Dimension 4: Samples of knowledge and skills relate to e-competences in dimension 2. They are provided to add value and context and are not intended to be exhaustive.

Competence definitions are explicitly assigned to dimensions 2 and 3 and knowledge and skills samples appear in dimension 4 of the framework; attitude is embedded in all three dimensions.

e-competences in dimensions 1 and 2 are presented from the organisational perspective as opposed to an individual's perspective. Figure 2.2 illustrates the ICT process stages as they are defined in the e-CF3.0 document.

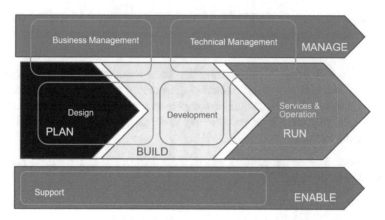

Fig. 2.2 ICT process stages aligned with the organisational production workflow (as used in e-CF3.0)

Table 2.1 contains competences defined for areas A–E. For more detailed definition of e-CF3.0 dimensions 1–3 and dimension 4, refer to Chap. 4 "Model Curriculum".

Figure 2.3 illustrates the multipurpose use of the European e-Competence Framework within ICT organisations. The e-CF has a multidimensional structure and is flexible in use for different purposes; it can be easily adopted for organisation-specific model and roles. The e-CF3.0 is used for job profiles definition in the European Committee for Standardization, CEN, Workshop Agreement, CWA, document 16458; see [11] and Chap. 5, Data Science Professional Profiles, which are linked to the organisational processes which creates limitations for cross-organisational professional profiles and roles such as data scientist. However, combining competences from different competence areas and using them as building blocks can allow flexible job profiles definition. This enables the derived job profiles to be easily updated by changing set of competences related to profiles without the need to restructure the entire profile.

The CF-DS development follows the European e-Competences Framework; although the e-CF3.0 provided a general framework for ICT competences definition and possible mapping to data science competences, it appeared that current e-CF3.0 does not contain competences that reflect specific data scientist role in organisation. Furthermore, e-CF3.0 is built around organisational workflow, while anticipated data scientist's role is cross-organisational bridging different organisational roles and departments in providing data-centric view or organisational processes.

On the other side European ICT profiles and its mapping to e-CF3.0 provided a good illustration of how individual ICT profiles can be mapped to e-CF3.0 competences and areas. Similarly, the additional ICT profiles are proposed to reflect data scientist's role in the organisation.

And finally, related to the EU, the European Skills, Competences, Occupations and Qualifications (ESCO) [7] provides a good example of a standardised

Table 2.1 e-CF3.0 competences defined for areas A-E

Dimension 1 :5 e-CF areas (A–E)	Dimension 2: 40 e-competences identified
A. Plan	A.1. IS and business strategy alignment
	A.2. Service level management
	A.3. Business plan development
	A.4. Product/service planning
	A.5. Architecture design
	A.6. Application design
	A.7. Technology trend monitoring
	A.8. Sustainable development
	A.9. Innovating
B. Build	B.1. Application development
	B.2. Component integration
	B.3. Testing
	B.4. Solution deployment
	B.5. Documentation production
	B.6. Systems engineering
C. Run	C.1. User support
	C.2. Change support
	C.3. Service delivery
	C.4. Problem management
D. Enable	D.1. Information security strategy development
	D.2. ICT quality strategy development
	D.3. Education and training provision
	D.4. Purchasing
	D.5. Sales proposal development
	D.6. Channel management
	D.7. Sales management
	D.8. Contract management
	D.9. Personnel development
	D.10. Information and knowledge management
	D.11. Needs identification
	D.12. Digital marketing
E. Manage	E.1. Forecast development
	E.2. Project and portfolio management
	E.3. Risk management
	E.4. Relationship management
	E.5. Process improvement
	E.6. ICT quality management
	E.7. Business change management
	E.8. Information security management
	E.9. IS governance

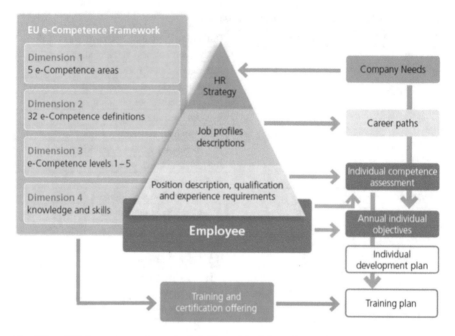

Fig. 2.3 e-CF3.0 structure and use for definition of the job profiles definition and training needs

competences and skills taxonomy. The presented study will provide contribution to the definition of the data scientist as new profession or occupation with related competences, skills and qualifications definition. The CF-DS definition will reuse, extend and map the ESCO taxonomy to the identified data science competences and skills.

2.1.3 Information Technology Competency Model, ACM IT-CM

The Association for Computing Machinery's (ACM) Information Technology Competency Model (IT-CM) of Core Learning Outcomes and Assessment for Associate-Degree Curriculum has been developed by ACM Committee for Computing Education in Community Colleges (ACM CCECC) [13].

The ACM CCECC classification is supported by the web portal [14]. The portal provides related information, linking and mapping between different classification systems, in particular:

- ACM Computing Classification System 2012
- United States Department of Labor Information Technology Competency Model
- Bloom's Revised Taxonomy [15]

- Program metrics for computer science, information technologies and software engineering programmes.

ACM currently categorises the overarching discipline of computing into five defined sub-disciplines: *computer science, computer engineering, software engineering, information systems* and *information technology*. This report specifically focuses on information technology defined by the ACM CCECC as follows: *Information technology involves the design, implementation and maintenance of technology solutions and support for users of such systems. Associated curricula focus on crafting hardware and software solutions as applied to networks, security, client-server and mobile computing, web applications, multimedia resources, communications systems and the planning and management of the technology life cycle.*

The IT-CM document refers to the US Department of Labor Information Technology Competency Model [16] that was one of the sources that provided a foundation for the curricular guidance outlined in IT-CM report.

Competencies are used to define the learning outcome. In formulating assessment rubrics, the ACM CCECC uses a structured template comprised of three tiers: *emerging, developed* and *highly developed,* which can actually be mapped to the level of Bloom's verbs from the Lower Order Thinking Skills (LOTS) to the Higher Order Thinking Skills (HOTS) including *analysing* and *evaluating.*

The ACM Competencies Model provides the basis for the competency-based learning, that is instead of focusing on how much time students spend learning a particular topic or concept, Carnegie unit credit hour, the outcomes-based model assesses whether students have mastered the given competencies, namely the skills, abilities and knowledge.

The document defines 50 learning outcomes (that also define the body of knowledge) that represent core or foundational competencies that a student in any IT-related programme must demonstrate. Curricula for specific IT programmes (e.g. networking, programming, digital media and user support) will necessarily include additional coursework in one or more defined areas of study. The core IT learning outcomes are grouped into technical competency areas and workplace skills.

ACM Computing Classification System will be used as a basis to define the proposed Data Science Body of Knowledge, and extension to ACM CCS2012 will be provided to cover the identified knowledge and required academic subjects. Necessary contacts will be done with the ACM CCS body and corresponding ACM curriculum defining committees.

2.1.4   O'Reilly Strata Industry Research

O'Reilly Strata industry research [17] defines the four data scientist professional profiles and their mapping to the basic set of technology domains and competencies

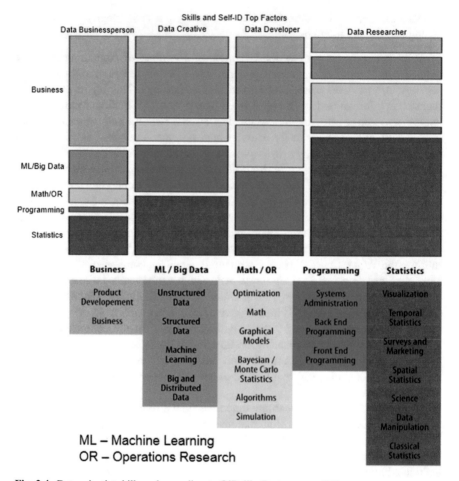

Fig. 2.4 Data scientist skills and according to O'Reilly Strata survey [17]

as shown in Fig. 2.4. The four profiles are defined based on the data science practitioners' self-identification:

- Data businessperson
- Data creative
- Data developer
- Data researcher

Table 2.2 lists skills for data science that are identified in the study. They are very specific in technical sense but provide useful information when mapped to the above-mentioned data science profiles. We will refer to this study in our analysis of CF-DS and related competence groups.

Table 2.2 Data scientist skills identified in the O'Reilly Strata study

Data science skills	Examples → Knowledge and skills
Algorithms	Computational complexity, CS theory
Back-end programming	JAVA/Rails/Objective C
Bayesian/Monte Carlo statistics	MCMC, BUGS
Big and distributed data	Hadoop, Map/Reduce
Business	Management, business development, budgeting
Classical statistics	General linear model, ANOVA
Data manipulation	Regexes, R, SAS, web scraping
Front-end programming	JavaScript, HTML, CSS
Graphical models	Social networks, Bayes networks
Machine learning	Decision trees, neural nets, SVM, clustering
Math	Linear algebra, real analysis, calculus
Optimisation	Linear, integer, convex, global
Product development	Design, project management
Science	Experimental design, technical writing/publishing
Simulation	Discrete, agent-based, continuous
Spatial statistics	Geographic covariates, GIS
Structured data	SQL, JSON, XML
Surveys and marketing	Multinomial modelling
Systems administration	*Nix, DBA, cloud tech.
Temporal statistics	Forecasting, time-series analysis
Unstructured data	NoSQL, text mining
Visualisation	Statistical graphics, mapping, web-based data visualisation

O'Reilly Strata study was a first extensive study on data scientist organisational roles, profiles and skills. Although skills are defined as very technically and technologically specific, the proposed definition of profiles is important for defining required competence groups; in particular identification of data science creative and data science researcher profiles indicates an important role of scientific approach and need for research method training in data scientist professional education. This group of competences is included in the proposed CF-DS.

2.1.5 Other Studies, Reports and Projects on Defining Data Science Competences

The following reports and studies and ongoing works to define the data science competences, skills and profiles and needs for European Research Area and industry are considered relevant to the current study and will be used to finalise the DS-CF definition:

- United Kingdom (UK) study of demand for big data skills. The study "Big Data Analytics: Assessment of demand for Labour and Skills 2013-2020" [18]

provided extensive analysis of the demand side for big data specialists in the UK in the forthcoming years. Although the majority of roles are identified as related to big data skills, it is obvious that all these roles can be related to more general definition of the data scientist as an organisational role working with big data and data-intensive technologies.

The report lists the following eight big data roles:

1. Big data developer
2. Big data architect
3. Big data analyst
4. Big data administrator
5. Big data consultant
6. Big data project manager
7. Big data designer
8. Data scientist

- Data Analytics Rising Employment (DARE) project is commissioned by Asia Pacific Economic Cooperation (APEC) council and is focused on defining the recommended data science analytics competences. The DARE project recommendation is to include the basic competences or literacy in the overall data science competences definition. In this project, the EDISON contributed to the definition of recommended data science analytics competences [19] and their alignment with the EDSF.
- LERU Roadmap for Research Data [20].
- ELIXIR community projects RITrain and CORBEL dealing with competences and skills definition for bioinformaticians as an example of data science-enabled professions.
- PwC and BHEF report "Investing in America's data science and analytics talent: The case for action" [21]
- Burning Glass Technology, IBM and BHEF report "The Quant Crunch: How the demand for Data Science Skills is disrupting the job Market" [22]

2.2 Definition of Data Science Competences, Skills and Knowledge

The competences definition has its strong foundation and roots in the existing frameworks treated in the previous section; for each one of them have been used standards and best practices documents that were used for defining the proposed set of data science competences and skills. The proposed CF-DS is defined as a three-dimensional competences model similar to e-CF3.0 (described in Sect. 2.1.2); the CF-DS uses a three-dimensional model that includes competences, skills and knowledge. Relation between competences, skills and knowledge is illustrated in Fig. 2.5. Competences ensure the ability to perform required organisational functions that are

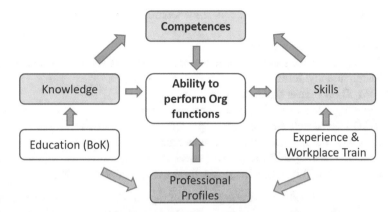

Fig. 2.5 Relation between competences, skills, knowledge and education

defined for specific organisational role that the worker/employee performs in the organisation. Competences must be supported byknowledge acquired in the process of education or training and by specific skills that is required as a result of practical activity or previous work in the similar role or profession. Knowledge and skills add to the ability and performance of organisational functions.

CF-DS adopts a holistic e-CF definition: *Competence is a demonstrated ability to apply knowledge, skills and attributes for achieving desirable results* in organisational or role context. CF-DS should work as an enabler for multiple applications that can be used by different types of users from individual to organisational; it should support common understanding and not mandate specific implementation.

2.2.1   3-Dimensional Competences Model

The results of the job market study and analysis for data science and data science-enabled vacancies, conducted at the initial stage of the project, provided the basis and justification for defining the main competence groups that are commonly required by companies, including identification skills such as data management and research methods that were not required formerly for data analytics jobs. In the following, the three dimensions of the CF-DS are explained.

2.2.1.1   CF-DS Competences

The following CF-DS five competence and skills groups have been identified:

1. **DSDA, Data analytics**

   Including statistical analysis, machine learning, data mining, business analytics, others
2. **DSENG, Data engineering**
   Including software and applications engineering, data warehousing, big data infrastructure and tools
3. **DSDM, Data management and governance**
   Including data stewardship, curation and preservation)
4. **DSRMP, Research methods and project management** for research-related professions and business process management for business-related p*rofessions*
5. **DSDK, Domain-specific knowledge and expertise** (Subject/scientific domain related). Also will be named, indistinctly, **business analytics, DSDA**

DSDA, DSENG and DSDM competence groups constitute the core data science competences that actually define the main Data Science Professional profiles and roles, including related to different application domains (see Sect. 2.2.6 discussion about relation between core data science concepts and competences and those related to knowledge and technology domains).

DSDM
DSDM and DSRMP competence groups are considered as commonly required for all Data Science Professional Profiles to ensure effective work with modern data-driven technologies and in modern data-driven organisations. Data management, curation and preservation competences are already attributed to the existing (research) data-related professions such as data stewards, data manager, data librarian, data archivist and others. Data management is an important component of European research area and open data and open access policies. It is extensively addressed by the Research Data Alliance (RDA) [4] and supported by numerous projects, initiatives and training programmes.

DSRMP
Knowledge of the research methods and techniques is something that makes data scientist profession different from all previous professions. It should be also coupled with the basic project management competences and skills.

   The research methods typically include the following stages:

- Design experiment
- Collect data
- Analyse data
- Identify patterns
- Hypothesise explanation
- Test hypothesis

Figure 2.6a, b provides graphical presentation of relations between identified competence groups as linked to research methods or to business process management. The figure illustrates importance of the data management competences and

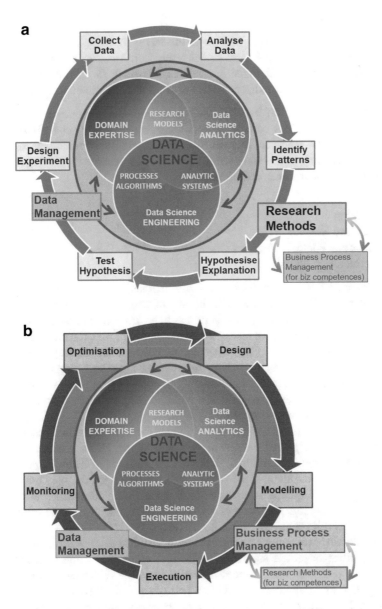

Fig. 2.6 Relations between identified data science competence groups. (**a**) Data science competences for general or research-oriented profiles. (**b**) Data science competences for business-oriented profiles

skills and research methods or business process management knowledge for all categories and profiles of data scientists.

DSDK (DSBA)

Important part of the research process is the theory building, but this activity is attributed to the domain or subject matter researcher. The data scientist (or related role) should be aware about domain-related research methods and theory as a part of their domain-related knowledge and team or workplace communications.

There is a number of the business process operations models depending on their purpose but typically they contain the following stages that are generally similar to those for scientific methods, in particular in collecting and processing data (see reference to existing definitions; see the Annex for reference to existing business process management stages definitions):

- Design
- Model/plan
- Deploy and execute
- Monitor and control
- Optimise and re-design

The identified demand for general competences and knowledge on data management and research methods needs to be implemented in the future data science education and training programmes, as well as to be included into re-skilling training programmes. It is important to mention that knowledge of research methods does not mean that all data scientists must be talented scientists; however, they need to know the general research methods such as formulating hypothesis, applying research methods, producing artefacts and evaluating hypothesis (so-called 4 steps model). Research methods training are already included into master's programmes and graduate students of many master's programmes.

From the education and training point of view, the identified competences can be treated or linked to expected learning or training outcome. This aspect is discussed in detail in relation to the definition of the Data Science Body of Knowledge and Data Science Model Curriculum in the next chapters. The identified five data science-related competence groups provide the basis for defining consistent and balanced education and training programmes for data science-related jobs, re-skilling and professional certification.

The following subsections provide the proposed data science competences definition for different groups supported by the data extracted for the collected information. The presented competences definition has been reviewed by a number of expert groups and individual experts as a part of the project EDISON engagement and network activities. The presented competences are required for different professional profiles, organisational roles and throughout the whole data life cycle, but not necessary to be provided by a single role or individual. The presented competences are enumerated to allow easy use and linking between the parts of the data science framework: CF-DS, DS-BoK, MC-DS and DSPP.

2.2.1.2   CF-DS Skills

The identified skills can be organised in the following two groups:

- **Group A Skills**: They refer to data science skills related to the main competence groups that cover knowledge and experience related to effectively realise defined competences and related organisational functions. The identified data science skills related to the main competence groups are:
  - Data analytics covering extensive skills related to using different machine learning, data mining, statistical methods and algorithms
  - Data engineering skills related to design, implementation and operation of the data science (or big data) infrastructure, platforms and applications
  - Data management and governance skills (including both general data management and research data management)
  - Research methods and project management skills
  - Business analytics as an example of domain-related skills

- The data analytics and data engineering groups are the most populated which reflects a wide spectrum of required skills in these groups as a core for the data science, because it is mandatory for the data scientist to have the ability to implement the effective data analytics solutions and applications.
- It is important to mention that the whole complex of data science-related competences, skills and knowledge are strongly based on the mathematical foundation that should include knowledge of mathematics, including linear algebra, calculus, statistics, probability theory and other mathematical subjects.
- **Group B Skills**: they refer to data analytics and data handling languages, tools, platforms and applications, including SQL- and NoSQL-based applications and data management tools and knowledge and experience with the big data infrastructure platforms and tools. The analysis and identified data science Group B skills are described in Sect. 2.2.6.
- **Transversal skills**: Separately defined are personal and attitude skills also referred to as transversal, the twenty-first-century skills and data science professional skills that define specific (personal) skills that the data scientist needs to develop to successfully work as a data scientist in different organisational roles and along their career. The analysis and identified data science soft skills are described in Sect. 2.2.6.

2.2.1.3   Knowledge Topics

Knowledge or knowledge topics or units are the required knowledge to support corresponding competence groups. There is no direct mapping between individual competences and knowledge topics; single competence may be mapped to multiple knowledge topics and vice versa. CF-DS provides mapping between knowledge topics defined for individual competences and knowledge units defined in DS-BoK defined in the next chapter.

2.2.2   Data Analytics Competences, Skills and Knowledge (DSDA)

In this section, the data analytics, DSDA, competences, skills group A and knowledge are presented.

Competences
Data analytics competences deal with the use of appropriate data analytics and statistical techniques on available data to discover new relations and deliver insights into research problem or organisational processes and support decision-making and cover extensive skills related to using different machine learning, data mining, statistical methods and algorithms.

The following are the six DSDA identified competences:

1. DSDA01. Effectively use variety of data analytics techniques, such as machine learning (including supervised, unsupervised, semi-supervised learning), data mining, prescriptive and predictive analytics, for complex data analysis through the whole data life cycle
2. DSDA02. Apply designated quantitative techniques, including statistics, time series analysis, optimisation and simulation to deploy appropriate models for analysis and prediction
3. DSDA03. Identify, extract and pull together available and pertinent heterogeneous data, including modern data sources such as social media data, open data, governmental data, verify data quality
4. DSDA04. Understand and use different performance and accuracy metrics for model validation in analytics projects, hypothesis testing and information retrieval
5. DSDA05. Develop required data analytics for organisational tasks, integrate data analytics and processing applications into organisation workflow and business processes to enable agile decision-making
6. DSDA06. Visualise results of data analysis, design dashboard and use storytelling methods

Skills Group A
The following are the sixteen data science and data analytics identified Group A skills, SDSDA:

1. SDSDA01. Use machine learning technology, algorithms, tools, including supervised, unsupervised or reinforced learning
2. SDSDA02. Use data mining techniques
3. SDSDA03. Use text data mining techniques

4. SDSDA04. General statistical analysis methods and techniques, descriptive analytics
5. SDSDA05. Use quantitative analytics methods
6. SDSDA06. Use qualitative analytics methods
7. SDSDA07. Apply predictive analytics methods
8. SDSDA08. Apply prescriptive analytics methods
9. SDSDA09. Use graph data analytics for organisational network analysis, customer relations, other tasks
10. SDSDA10. Apply analytics and statistics methods for data preparation and preprocessing
11. SDSDA11. Be able to use performance and accuracy metrics for data analytics assessment and validation
12. SDSDA12. Use effective visualisation and storytelling methods to create dashboards and data analytics reports
13. SDSDA13. Use natural language processing methods
14. SDSDA14. Operations research
15. SDSDA15. Optimisation
16. SDSDA16. Simulation

Knowledge Topics
The following are the eighteen data science and data analytics knowledge (KDSDA) required to support the identified competences in this subject:

1. KDSDA01. Machine learning supervised: decision trees, naïve Bayes classification, ordinary least square regression, logistic regression, neural networks, SVM (support vector machine), ensemble methods, others
2. KDSDA02. Machine learning unsupervised: clustering algorithms, principal component analysis (PCA), singular value decomposition (SVD), independent component analysis (ICA)
3. KDSDA03. Machine learning (reinforced): Q-learning, TD-learning, genetic algorithms)
4. KDSDA04. Data mining (text mining, anomaly detection, regression, time series, classification, feature selection, association, clustering)
5. KDSDA05. Text data mining: statistical methods, NLP, feature selection, a priori algorithm, etc.
6. KDSDA06. General statistical analysis methods and techniques, descriptive analytics
7. KDSDA07. Quantitative analytics
8. KDSDA08. Qualitative analytics
9. KDSDA09. Predictive analytics
10. KDSDA10. Prescriptive analytics
11. KDSDA11. Graph data analytics: path analysis, connectivity analysis, community analysis, centrality analysis, sub-graph isomorphism, etc.
12. KDSDA12. Natural language processing
13. KDSDA13. Data preparation and preprocessing

14. KDSDA14. Performance and accuracy metrics
15. KDSDA15. Markov models, conditional random fields
16. KDSDA16. Operations research
17. KDSDA17. Optimisation
18. KDSDA18. Simulation

2.2.3   Data Engineering Competences, Skills and Knowledge (DSDEG)

In this section, the data science engineering, DSENG, competences, Group A skills and knowledge are presented.

Competences

Data science engineering competences deal with the use of engineering principles and modern computer technologies to research, design, implement new data analytics applications; develop experiments, processes, instruments, systems, infrastructures to support data handling during the whole data life cycle.

The following are the six DSENG identified competences:

1. DSENG01. Use engineering principles (general and software) to research, design, develop and implement new instruments and applications for data collection, storage, analysis and visualisation
2. DSENG02. Develop and apply computational and data-driven solutions to domain-related problems using a wide range of data analytics platforms, with the special focus on big data technologies for large datasets and cloud-based data analytics platforms
3. DSENG03. Develop and prototype specialised data analysis applications, tools and supporting infrastructures for data-driven scientific, business or organisational workflow; use distributed, parallel, batch and streaming processing platforms, including online and cloud-based solutions for on-demand provisioned and scalable services
4. DSENG04. Develop, deploy and operate large-scale data storage and processing solutions using different distributed and cloud-based platforms for storing data (e.g. Data Lakes, Hadoop, HBase, Cassandra, MongoDB, Accumulo, DynamoDB, others)
5. DSENG05. Consistently apply data security mechanisms and controls at each stage of the data processing, including data anonymisation, privacy and IPR protection
6. DSENG06. Design, build, operate relational and non-relational databases (SQL and NoSQL), integrate them with the modern data warehouse solutions, ensure effective ETL (extract, transform, load), OLTP, OLAP processes for large datasets

Skills Group A

The following are the 12 data science engineering identified Group A skills, SDSENG:

1. SDSENG01. Use systems and software engineering principles to organisations information system design and development, including requirements design
2. SDSENG02. Use cloud computing technologies and cloud-powered services design for data infrastructure and data handling services
3. SDSENG03. Use cloud-based big data technologies for large datasets processing systems and applications
4. SDSENG04. Use agile development technologies, such as DevOps and continuous improvement cycle, for data-driven applications
5. SDSENG05. Develop and implement systems and data security, data access, including data anonymisation, federated access control systems
6. SDSENG06. Apply compliance-based security models, in particular for privacy and IPR protection
7. SDSENG07. Use relational, non-relational databases (SQL and NoSQL), data warehouse solutions, ETL (extract, transform, load), OLTP, OLAP processes for structured and unstructured data
8. SDSENG08. Effectively use big data infrastructures, high-performance networks, infrastructure and services management and operation
9. SDSENG09. Use and apply modelling and simulation technologies and systems
10. SDSENG10. Use and integrate with the organisational information systems, collaborative system
11. SDSENG11. Design efficient algorithms for accessing and analysing large amounts of data, including API to different databases and datasets
12. SDSENG12. Use of recommender or ranking system

Knowledge Topics

The following are the 22 data science engineering knowledge (KDSENG) required to support the identified competences in this subject:

1. KDSENG01. Systems engineering and software engineering principles, methods and models, distributed systems design and organisation
2. KDSENG02. Cloud computing, cloud-based services and cloud-powered services design
3. KDSENG03. Big data technologies for large datasets processing: batch, parallel, streaming systems, in particular cloud based
4. KDSENG04. Applications software requirements engineering and design, agile development technologies, DevOps and continuous improvement cycle
5. KDSENG05. Systems and data security, data access, including data anonymisation, federated access control systems
6. KDSENG06. Compliance-based security models, privacy and IPR protection

7. KDSENG07. Relational, non-relational databases (SQL and NoSQL), data warehouse solutions, ETL (extract, transform, load), OLTP, OLAP processes for large datasets
8. KDSENG08. Big data infrastructures, high-performance networks, infrastructure and services management and operation
9. KDSENG09. Modelling and simulation, theory and systems
10. KDSENG10. Information systems, collaborative systems

2.2.4   Data Management Competences, Skills and Knowledge (DSDM)

In this section, the data science data management and governance, DSDM, competences, Group A skills and knowledge are presented.

Competences
Data science management and governance competences deal with develop and implement data management strategy for data collection, storage, preservation and availability for further processing.

The following are the six DSDM identified competences:

1. DSDM01. Develop and implement data strategy, in particular, in the form of data management policy and data management plan (DMP)
2. DSDM02. Develop and implement relevant data models, define metadata using common standards and practices, for different data sources in a variety of scientific and industry domains
3. DSDM03. Integrate heterogeneous data from multiple source and provide them for further analysis and use
4. DSDM04. Maintain historical information on data handling, including reference to published data and corresponding data sources (data provenance)
5. DSDM05. Ensure data quality, accessibility, interoperability, compliance to standards and publication (data curation)
6. DSDM06. Develop and manage/supervise policies on data protection, privacy, intellectual property Rights, IPR and ethical issues in data management

Skills Group A
The following are the nine data management and governance Group A skills, SDSDM:

1. SDSDM01. Specify, develop and implement enterprise data management and data governance strategy and architecture, including data management plan (DMP)
2. SDSDM02. Data storage systems, data archive services, digital libraries and their operational models

3. SDSDM03. Define requirements to and supervise implementation of the hybrid data management infrastructure, including enterprise private and public cloud resources and services
4. SDSDM04. Develop and implement data architecture, data types and data formats, data modelling and design, including related technologies (ETL, OLAP, OLTP, etc.)
5. SDSDM05. Implement data lifecycle support in organisational workflow, support data provenance and linked data
6. SDSDM06. Consistently implement data curation and data quality controls, ensure data integration and interoperability
7. SDSDM07. Implement data protection, backup, privacy, mechanisms/services, comply with IPR, ethics and responsible data use
8. SDSDM08. Use and implement metadata, PID, data registries, data factories, standards and compliance
9. SDSDM09. Adhere to the principles of the open data, open science, open access, use ORCID-based services

Knowledge Topics
The following are the nine data management and governance knowledge, KDSDM, required to support the identified competences in this subject:

1. KDSDM01. Data management and enterprise data infrastructure, private and public data storage systems and services
2. KDSDM02. Data storage systems, data archive services, digital libraries and their operational models
3. KDSDM03. Data governance, data governance strategy, data management plan (DMP)
4. KDSDM04. Data architecture, data types and data formats, data modelling and design, including related technologies (ETL, OLAP, OLTP, etc.)
5. KDSDM05. Data lifecycle and organisational workflow, data provenance and linked data
6. KDSDM06. Data curation and data quality, data integration and interoperability
7. KDSDM07. Data protection, backup, privacy, IPR, ethics and responsible data use
8. KDSDM08. Metadata, PID, data registries, data factories, standards and compliance
9. KDSDM09. Open data, open science, research data archives/repositories, open access, ORCID

2.2.5 Research Methods and Project Management Competences, Skills and Knowledge, DSRPM

In this section, the research methods and project management (DSRPM) competences, Group A skills and knowledge are presented.

Competences
Data science research methods and project management competences create new understandings and capabilities by using the scientific method (hypothesis, test/ artefact, evaluation) or similar engineering methods to discover new approaches to create new knowledge and achieve research or organisational goals.

The following are the six DSRMP identified competences:

1. DSRMP01. Create new understandings by using the research methods (including hypothesis, artefact/experiment, evaluation) or similar engineering research and development methods
2. DSRMP02. Direct systematic study towards understanding of the observable facts, and discover new approaches to achieve research or organisational goals
3. DSRMP03. Analyse domain-related research process model, identify and analyse available data to identify research questions and/or organisational objectives and formulate sound hypothesis
4. DSRMP04. Undertake creative work, making systematic use of investigation or experimentation, to discover or revise knowledge of reality, and use this knowledge to devise new applications, contribute to the development of organisational objectives
5. DSRMP05. Design experiments which include data collection (passive and active) for hypothesis testing and problem-solving
6. DSRMP06. Develop and guide data-driven projects, including project planning, experiment design, data collection and handling.

Group A Skills
The following are the six research methods and project management Group A skills, SDSRMP:

1. SDSRMP01. Use research methods principles in developing data-driven applications and implementing the whole cycle of data handling
2. SDSRMP02. Design experiment, develop and implement data collection process
3. SDSRMP03. Apply data lifecycle management model to data collection and data quality evaluation
4. SDSRMP04. Apply structured approach to use case analysis
5. SDSRMP05. Develop and implement research data management plan (DMP), apply data stewardship procedures
6. SDSRMP06. Consistently apply project management workflow: scope, planning, assessment, quality and risk management, team management

Knowledge Topics
The following are the six research methods and project Management knowledge, KDSRMP, required to support the identified competences in this subject:

1. KDSRMP01. Research methods, research cycle, hypothesis definition and testing
2. KDSRMP02. Experiment design, modelling and planning
3. KDSRMP03. Data life cycle and data collection, data quality evaluation
4. KDSRMP04. Use case analysis: research infrastructure and projects
5. KDSRMP05. Research data management plan (DMP) and data stewardship
6. KDSRMP05. Research data management plan (DMP) and data stewardship
7. KDSRMP06. Project management: scope, planning, assessment, quality and risk management, team management

2.2.6 Domain-Specific Competences Group (DSDK)

Domain knowledge and expertise (DSDK) competences deal with the domain knowledge (scientific or business) to develop relevant data analytics applications; adopt general data science methods to domain-specific data types and presentations, data and process models, organisational roles and relations.

Data scientist, by definition, is playing assistant role to the main organisational management role at the level of decision-making or a subject domain scientific/ researcher role to help them with organising data management and data processing to achieve their specific management or research role. However, data scientist has also an opportunity to play a leading role in some data-driven projects or functions because of their potentially wider vision of the organisational processes or influencing factors.

To understand this, we need to look closer at relation between data scientist and subject domain specialist. The subject domain is generally defined by the following components:

- Model (and data types)
- Methods (and additionally theory)
- Processes
- Domain-specific data types and presentation, including visualisation methods
- Organisational roles and relations

Data scientist as an assistant to the subject domain specialist will do the following work that should bring benefits to organisation or facilitate scientific discovery:

- Translate subject domain model, methods, processes into abstract data-driven form
- Implement computational models in software, build required infrastructure and tools
- Do (computational) analytic work and present it in a form understandable to subject domain

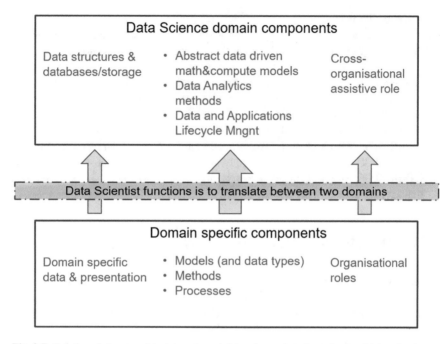

Fig. 2.7 Relations between subject domain and data science domain and role of data scientist

- Discover new relations originated from data analysis and advise subject domain specialist
- Interact and cooperate with different organisational roles to obtain data and deliver results and/or actionable data

Formalisation of the relations between the components and work activities of the subject domain specialist/scientist and data science domain provides additional arguments to the discussion about the data scientist contribution to the scientific research and discovery that has been recently disputed in many forums: Should data scientist be treated as an author of the potential scientific discovery, or just be acknowledged for contribution as assistant role?

Figure 2.7 illustrates relations between subject domain components and those mapped to data science domain which is abstract, formalised and data driven.

This section provides example definitions of the domain-specific knowledge and expertise (DSDK) or business analytics (DSBA) competence group.

Competences

The following are the six DSBA (or DSDK) identified competences:

1. DSBA01. Analyse information needs, assess existing data and suggest/identify new data required for specific business context to achieve organisational goal, including using social network and open data sources

2. DSBA02. Operationalise fuzzy concepts to enable key performance indicators measurement to validate the business analysis, identify and assess potential challenges
3. DSBA03. Deliver business focused analysis using appropriate BA/BI methods and tools, identify business impact from trends; make business case as a result of organisational data analysis and identified trends
4. DSBA04. Analyse opportunity and suggest use of historical data available at organisation for organisational processes optimisation
5. DSBA05. Analyse customer relations data to optimise/improve interacting with the specific user groups or in the specific business sectors
6. DSBA06. Analyse multiple data sources for marketing purposes; identify effective marketing actions

Group A Skills
The following are the nine domain knowledge and expertise Group A skills (SDSBA):

1. SDSBA01. Business Intelligence (BI) methods for data analysis; apply cognitive technologies and relevant services
2. SDSBA02. Apply Business Processes Management (BPM), general business processes and operations for organisational processes analysis/modelling
3. SDSBA03. Apply agile data-driven methodologies, processes and enterprises
4. SDSBA04. Use econometrics for data analysis and applications
5. SDSBA05. Develop data-driven customer relations management (CRP), user experience (UX) requirements and design
6. SDSBA06. Apply structured approach to use case analysis in business and industry
7. SDSBA07. Use data warehouses technologies for data integration and analytics, including use open data and social media data
8. SDSBA08. Use data-driven marketing technologies
9. SDSBA09. Mechanism design and/or latent Dirichlet allocation

Knowledge
The following are the eight domain knowledge and expertise knowledge (KDSBA) required to support the identified competences in this subject:

1. KDSBA01. Business analytics (BA) and business intelligence (BI); methods and data analysis; cognitive technologies
2. KDSBA02. Business processes management (BPM), general business processes and operations, organisational processes analysis/modelling
3. KDSBA03. Agile data-driven methodologies, processes and enterprises
4. KDSBA04. Econometrics: data analysis and applications
5. KDSBA05. Data-driven customer relations management (CRP), user experience (UX) requirements and design
6. KDSBA06. Use case analysis: business and industry
7. KDSBA07. Data warehouses technologies, data integration and analytics

8. KDSBA08. Data-driven marketing technologies

2.3   Data Science Additional Skills

2.3.1   Group B Skills

Group B skills are common practical skills related to using computational and data management platforms, programming languages and tools. In Sect. 2.2 it has been said that Group B skills are all related to data analytics and data handling languages, tools, platforms and applications, including SQL- and NoSQL-based applications and data management tools and knowledge and experience with the big data infrastructure platforms and tools.

The identified skills related to the data analytics languages, tools, platforms and big data infrastructure that are split into six subgroups. The groups and their associated skills are:

1. Data analytics and statistical languages and tools.
    The following are the twelve data analytics and statistical languages and tools Group B skills, DSDALANG:

    1.1. DSDALANG01. R and data analytics libraries (CRAN, ggplot2, dplyr, reshap2, etc.)
    1.2. DSDALANG02. Python and data analytics libraries (pandas, numpy, mathplotlib, scipy, scikit-learn, seaborn, beautifulsoup4, etc.)
    1.3. DSDALANG03. SAS
    1.4. DSDALANG04. IBM SPSS
    1.5. DSDALANG05. Julia
    1.6. DSDALANG06. RapidMiner
    1.7. DSDALANG07. Other analytics, statistical and programming languages (WEKA, KNIME, Scala, Stata, Orange, etc.)
    1.8. DSDALANG08. Scripting language, e.g. Octave, PHP, Pig, HiveQL, others
    1.9. DSDALANG09. Matlab data analytics
    1.10. DSDALANG10. Analytics tools (R/R Studio, Python/Anaconda, SPSS, Matlab, etc.)
    1.11. DSDALANG11. Data mining tools: RapidMiner, Orange, R, WEKA, NLTK, others
    1.12. DSDALANG12.   Excel   data   analytics   (Analysis   ToolPack, PivotTables, etc.)

2. Databases and query languages
    The following are the five databases and query languages Group B skills (DSADB):

    2.1. DSADB01. SQL and relational databases (incl. open source: PostgreSQL, mySQL, Nettezza, etc.)

2.2. DSADB02. SQL and relational databases (proprietary: Oracle, MS SQL Server, others)

2.3. DSADB03. NoSQL Databases (HBase, Hive, MongoDB, Cassandra, Redis, Accumulo, etc.)

2.4. DSADB 04. Hive (query language for Hadoop)

2.5. DSADB 05. Data modelling (UML, ERWin, DDL, etc.)

3. Data/applications visualisation

The following are the three data/applications visualisation Group B skills, DSVIZ:

3.1. DSVIZ01. Data visualisation libraries (mathpoltlib, seaborn, D3.js, FusionCharts, Chart.js, others)

3.2. DSVIZ02. Visualisation software (D3.js, Processing, Tableau, Raphael, Gephi, etc.)

3.3. DSVIZ03. Online visualisation tools (Datawrapper, Google Visualisation API, Google Charts, Flare, etc.)

4. Data management and curation platform

The following are the five data management and curation platform Group B skills (DSADM):

4.1. DSADM01. Data modelling and related technologies (ETL, OLAP, OLTP, etc.)

4.2. DSADM02. Data warehouse platform and related tools

4.3. DSADM03. Data curation platform, metadata management (ETL, Curator's Workbench, DataUp, MIXED, etc.)

4.4. DSADM04. Backup and storage management (iRODS, XArch, Nesstar, others)

4.5. DSADM05. Big data and cloud-based storage platforms and services

5. Big data analytics platforms

The following are the eleven big data analytics platforms Group B skills (DSBDA):

5.1. DSBDA01. Big data and distributed computing tools (Hadoop, Spark, MapReduce, Mahout, Lucene, NLTK, Pregel, etc.)

5.2. DSBDA02. Big data analytics platforms (Hadoop, Spark, Data Lakes, others)

5.3. DSBDA03. Real-time and streaming analytics systems (Flume, Kafka, Storm)

5.4. DSBDA04. Hadoop ecosystem/platform (Apache, Cloudera, Hortonworks, etc.)

5.5. DSBDA05. Azure data analytics platforms (HDInsight, Data Lake Analytics, PowerBI, Machine Learning Studio, etc.)

5.6. DSBDA06. Amazon Data Analytics platform (EMR, Kinesis, Data Pipeline, Machine Learning, etc.)

5.7. DSBDA07. Google Analytics platform (Google Data Studio, Machine Learning, TensorFlow, others)

5.8. DSBDA08. IBM Watson Analytics

5.9. DSBDA09. Other cloud-based data analytics platforms (Cloudera Data Science Workbench, HortonWorks Data Science, Vertica, LexisNexis HPCC System, etc.)

5.10. DSBDA10. Cognitive platforms (such as IBM Watson, Microsoft Cortana, others)

5.11. DSBDA11. Kaggle competition, resources and community platform

6. Development and project management frameworks, platforms and tools
   The following are the four development and project management frameworks, platforms and tools Group B skills (DSDEV):

6.1. DSDEV01. Frameworks: Python, Java, C/C++, GO, D3.js (data-driven documents), jQuery, others

6.2. DSDEV02. Python, Java or C/C++ development platforms/IDE (Eclipse, R Studio, Anaconda/Jupyter Notebook, Visual Studio Code, Atom, others)

6.3. DSDEV03. Git versioning system as a general platform for software development

6.4. DSDEV04. Scrum agile software development and management methodology and platform

It is also important for data scientist to be familiar with multiple data analytics languages and demonstrate proficiency in one or few the most popular languages (what should be supported with several years of practical experience), such as:

- R including extensive data analysis libraries
- Python and related data analytics libraries
- Julia
- SPSS
- KNIME, Orange, WEKA, others

Data science practitioner must be familiar and have experience with the general programming languages, software versioning and project management environments such as:

- Java, JavaScript and/or C/C++ as general applications programming languages
- Git versioning system as a general platform for software development
- Scrum agile software development and management methodology and platform

It is essential to mention that all modern big data platforms and general data storage and management platforms are cloud based. The knowledge of cloud computing and related platforms for applications deployment and data management are included in the table. The use of cloud-based data analytics tools is growing and most of big cloud services providers provide whole suites of platforms and tools for enterprise data management from enterprise data warehouses, data backup and archiving to business data analytics, data visualisation and content streaming.

2.3.2 Transversal Skills

Although it is commonly agreed on the importance of the soft skills for data scientist, the job market analysis clearly confirmed the importance of personal skills and identified a number of specific data science professional skills that are required for the data scientist to effectively work in the modern agile data-driven organisations and project teams. These should be also complemented with the general personal skills referred to as twenty-first century skills. Importance of such skills for data scientist is defined by their cross-organisational functions and responsibilities in collecting and analysing organisational data to provide insight for decision-making. In such a role the data scientist often reports to executive level or to other departments and teams. These skills extend beyond traditionally required communication or team skills. In addition, the ideal data scientist is expected to bring and spread new knowledge to organisation and ensure that all benefit and contribute to the processes related to data collection, analysis and exploitation.

In consequence, the main two transversal skills and their associated skills are:

2.3.2.1 Data Science Professional or Attitude Skills (DSPS) (Thinking and Acting Like a Data Scientist)

The data science is growing as a distinct profession and consequently will need professional identification via definition of the specific professional skills and code of conduct that can be defined as "Thinking and acting like Data Scientist". Understanding, recognising and acquiring such skills is essential for the data scientist to successfully progress along their career. It is also important for team leaders to correctly build relations in the team of project group.

Below are listed the data science professional (or attitude) skills (DSPS) which are identified by the data science practitioners and educators. Although some of the skills are common to the twenty-first-century skills, it is important to provide the whole list of skills that can provide guidance for future data scientists what skills are expected from them and need to be developed along their career.

- DSPS01. Accept/be ready for iterative development, know when to stop, comfortable with failure, accept the symmetry of outcome (both positive and negative results are valuable)
- DSPS02. Ask the right questions
- DSPS03. Recognise what things are important and what things are not important
- DSPS04. Respect domain/subject matter knowledge in the area of data science
- DSPS05. Data-driven problem solver and impact-driven mindset
- DSPS06. Recognise value of data, work with raw data, exercise good data intuition
- DSPS07. Good sense of metrics, understand importance of the results validation, never stop looking at individual examples

- DSPS08. Be aware about power and limitations of the main machine learning and data analytics algorithms and tools
- DSPS09. Understand that most of data analytics algorithms are statistics and probability based, so any answer or solution has some degree of probability and represents an optimal solution for a number of variables and factors
- DSPS10. Working in agile environment and coordinate with other roles and team members
- DSPS11. Work in multidisciplinary team, ability to communicate with the domain and subject matter experts
- DSPS12. Embrace online learning, continuously improve your knowledge, use professional networks and communities
- DSPS13. Storytelling: Deliver actionable result of your analysis
- DSPS14. Attitude: Creativity, curiosity (willingness to challenge status quo), commitment in finding new knowledge and progress to completion
- DSPS15. Ethics and responsible use of data and insight delivered, awareness of dependability (data scientist is a feedback loop in data driven companies).

2.3.2.2 Twenty-First-Century Skills (SK21) (Aka "Soft" Skills)

Twenty-first-century skills comprise a set of general workplace skills that include critical thinking, creativity, communication, collaboration, organisational awareness, ethics and others. The importance of this kind of skills is motivated by the fast technology development and the ongoing digital transformation of modern economy and Industry 4.0.

Below are listed the twenty-first-century skills (SK 21) defined based on the recommendations of the DARE Project [19], OECD Report on industry digitalisation [23] and P21's Framework for twenty-first-century learning [24].

- SK21C01. Critical Thinking: Demonstrating the ability to apply critical thinking skills to solve problems and make effective decisions
- SK21C02. Communication: Understanding and communicating ideas
- SK21C03. Collaboration: Working with other, appreciation of multicultural difference
- SK21C04. Creativity and attitude: Deliver high-quality work and focus on final result, initiative, intellectual risk
- SK21C05. Planning and organising: Planning and prioritising work to manage time effectively and accomplish assigned tasks
- SK21C06. Business fundamentals: Having fundamental knowledge of the organisation and the industry
- SK21C07. Customer focus: Actively look for ways to identify market demands and meet customer or client needs
- SK21C08. Working with tools and technology: Selecting, using and maintaining tools and technology to facilitate work activity

- SK21C09. Dynamic (self-) re-skilling: Continuously monitor individual knowledge and skills as shared responsibility between employer and employee, ability to adopt to changes
- SK21C10. Professional network: Involvement and contribution to professional network activities
- SK21C11. Ethics: Adhere to high ethical and professional norms, responsible use of power data-driven technologies, avoid and disregard unethical use of technologies and biased data collection and presentation.

Chapter 3
Data Science Body of Knowledge

Juan J. Cuadrado-Gallego and Yuri Demchenko

This chapter presents the definition of a consistent body of knowledge for data science (DS-BoK), which have three main objectives: (1) Support the competence groups defined in the Competences Framework for Data Science (CF-DS) presented in the previous chapter; (2) reflect the data-lifecycle management where different organisational roles, functions, competences and knowledge are required; and (3) ensure knowledge transferability and education programmes compatibility.

Extending the third objective, the DS-BoK can be used as the basis for defining data science-related curricula, courses, instructional methods, educational/course materials and necessary practices for university undergraduate and postgraduate programmes and professional training courses. The DS-BoK is also intended to be used for defining certification programmes and certification exam questions. While CF-DS (comprising of competences, skills and knowledge) can be used for defining job profiles (and correspondingly the content of job advertisements), the DS-BoK can provide the basis for interview questions and evaluation of the candidate's knowledge and related skills, as well as for professional certification exam and training.

This chapter, Data Science Body of Knowledge, DS-BoK, has three sections with the following structure: Sect. 3.1 presents an overview of related bodies of knowledge used to develop the DS-BoK; Sect. 3.2 presents the definition of the DS-BoK, with five subsections, one for each one of the five knowledge area groups defined for the DS-BoK; each subsection has a different number of subsections, one for each knowledge area identified for the knowledge area group; and finally Sect. 3.3

J. J. Cuadrado-Gallego (✉)
Department of Computer Science, University of Alcalá, Madrid, Spain
e-mail: jjcg@uah.es

Y. Demchenko
Universiteit van Amsterdam, Amsterdam, The Netherlands

© Springer Nature Switzerland AG 2020
J. J. Cuadrado-Gallego, Y. Demchenko (eds.), *The Data Science Framework*,
https://doi.org/10.1007/978-3-030-51023-7_3

presents the use of ACM-IEEE CCS2012 [14] in the DS-BoK and a proposed extension of ACM-IEEE CCS2012 from the DS-BoK.

3.1 Overview of Related Bodies of Knowledge

To develop the definition of the Data Science Body of Knowledge (DS-BoK), a previous analysis of existing bodies of knowledge, BoKs, was carried out. This analysis allowed us to identify what existing BoKs can be used in the DS-BoK definition, and which of their components could be reused to construct the DS-BoK. The proposed DS-BoK reuses where possible existing BoKs taking necessary Knowledge Area (KA) and Knowledge Unit (KU) definitions and combining them into above-defined DS-BoK knowledge area groups.

The BoKs that have been reviewed and used in the definition of the DS-BoK are:

- Computer Science Body of Knowledge (ACM-IEEE CS-BoK) [14, 25, 26]
- Information and Communication Technologies Professional Body of Knowledge (EC ICT-BoK) [27]
- Software Engineering Body of Knowledge (IEEE SWEBOK) [28]
- Business Analytics Body of Knowledge (IIBA BABOK) [29]
- Data Management Body of Knowledge (DAMA DMBOK) [30]
- Project Management Professional Body of Knowledge (PMI PM-BoK) [31]

Also, the Classification Computer Science CCS2012 for computer science-related knowledge areas was used, which will be treated in detail in Sect. 3.3.

From this initial analysis, the relevant best practices have been identified to structure the DS-BoK and provide the basis for defining the Data Science Body of Knowledge. This is also enriched by analysis of the practice in academic and professional training courses development by universities and professional training organisations.

It is also anticipated that due to the complexity of data science domain, the DS-BoK will require a wide spectrum of background knowledge, first of all in mathematics, statistics, logics and reasoning as well as general computing, and cloud computing in particular. Similar to the ACM CS2013 curricula approach, background knowledge can be required as an entry condition or must be studied as elective courses.

In the following six subsections, a brief introduction of each one of the BoKs used in the definition of the DS-BoK is presented, also each BoK has been used for developing different parts of the BoK and some examples of each use are described:

3.1.1 Computer Science Body of Knowledge (ACM-IEEE CS-BoK)

The Computer Science Body of Knowledge (CS-BoK) was published jointly by the Association for Computing Machinery (ACM) and the Institute of Electrical and Electronics Engineers (IEEE) as a part of the Computer Science Curricula 2013, ACM-IEEE CS2013 [25]. In this document, the body of knowledge in computer science is defined as a specification of the knowledge that must be covered in a curriculum in computer science.

The ACM-IEEE CS-BoK structures and defines the knowledge needed to define a curriculum in computer science in 18 knowledge areas (KA):

1. AL—Algorithms and complexity
2. AR—Architecture and organisation
3. CN—Computational science
4. DS—Discrete structures
5. GV—Graphics and visualisation
6. HCI—Human–computer interaction
7. IAS—Information assurance and security
8. IM—Information management
9. IS—Intelligent systems
10. NC—Networking and communications
11. OS—Operating systems
12. PBD—Platform-based development
13. PD—Parallel and distributed computing
14. PL—Programming languages
15. SDF—Software development fundamentals
16. SE—Software engineering
17. SF—Systems fundamentals
18. SP—Social issues and professional practice

Each knowledge area should not directly match a particular course in a curriculum but courses address topics from multiple knowledge areas.

Jointly with the KA, the ACM-IEEE CS2013 distinguishes between two types of topics:

- Core topics: Topics that must be taught to every student. They are subdivided into

  - Tier 1: Topics that are mandatory for each curriculum
  - Tier 2: Topics that are generally essential in an undergraduate computer science degree. It is expected that they be covered 90–100% with minimum advised 80% in each curriculum

- Elective topics: Topics that extend the breadth and depth of the knowledge in computer science and that have not been covered in the core.

Topics are defined differently for different programmes and specialisations. The reason for such a hierarchical approach to the structure of the body of knowledge is a useful way to group-related information, not as a structure for organising material into courses.

The structure of the description of each KA is the following: First a wide description of the KA is presented and then are listed and explained all the topics of the KA, indicating the core Tier 1 and 2, and the elective with the knowledge that must be studied in each core or elective and the learning outcomes of each one of them.

Other important feature of the ACM-IEEE CS-BoK is that it uses the 2012 ACM Computing Classification System (CCS) [14], which is widely accepted as the de facto standard classification system for the computing field.

Cs-BoK has been used, for example, in the data analytics knowledge area group, or in the data science engineering knowledge area group; in this some of the KAs used were algorithms and complexity, architecture and organisation (including computer architectures and network architectures), computational science, graphics and visualisation, information management, platform-based development, software engineering.

3.1.2 Foundational Information and Communication Technologies Body of Knowledge (EC ICT-BoK)

The Information and Communication Technologies Professional Body of knowledge (ICT-BoK) [27] was developed by Capgemini Consulting and Ernst & Young and published by the Directorate General Internal Market, Industry, Entrepreneurship and Small and Medium-sized Enterprises of the European Commission, EC, as part of the Information and Communication Technologies (ICT) Professionalism Framework developed in the e-Skills & Information and Communication Technologies Professionalism project carried out by the Directorate General Enterprise and Industry of the European Commission, which is formed by four building blocks: body of knowledge (BoK); competence framework; education and training resources; and code of professional ethics. In this document, the body of knowledge in ICT is defined as a specification of the base-level knowledge required to enter the ICT profession and acts as the first point of reference for anyone interested in working in ICT.

The EC ICT-BoK structures and defines the knowledge needed in information and communication technologies in 12 knowledge areas (KA):

1. ICT strategy and governance
2. Business and market of ICT
3. Project management
4. Security management
5. Quality management

   6. Architecture
   7. Data and information management
   8. Network and systems integration
   9. Software design and development
  10. Human–computer interaction
  11. Testing
  12. Operations and service management

The contents of each KA are described in depth in Sect. 3 *Building blocks of the Foundational ICT Body of Knowledge* of the EC ICT-BoK. The structure of the description of each KA is the following:

- First a wide description of the KA is presented.
- Then all the items required as foundational knowledge of the KA are listed.
- Then the references to dimension 4: knowledge of the e-Competence Framework for the KA are listed.
- Then examples of possible job profiles that require understanding the knowledge area are listed.
- Finally, examples of specific bodies of knowledge, certification and training possibilities related with that KA are listed.

In addition, the EC ICT-BoK has a Sect. 3.2, called *Introduction to Cross-cutting Knowledge Areas* in which are presented three knowledge dimensions that are cross-cutting and that can be applied to each of the 12 knowledge areas described above. The three new knowledge dimensions are:

  13. Soft skills
  14. IT legal, ethical, social and professional practices
  15. Emerging and disruptive technologies

The contents of each of these new KAs are described in depth in the same Sect. 3.2. The structure of the description of each KA is the following: First a wide description of the KA is presented; then all the items required as foundational knowledge of the KA are listed; then the references to existing bodies of knowledge are listed; and finally the references to dimension 4: knowledge of the e-Competence Framework for the KA are listed.

Among the proposed uses by the European Commission of ICT-BoK are *to be a source of inspiration for curricula design and development* and *promote the body of knowledge to their members, ICT professionals*. Taking into account this use of the ICT-BoK, for example, to develop data science engineering knowledge area group, the KAs used were project management, security management and quality management.

3.1.3 Software Engineering Body of Knowledge (IEEE SWEBOK)

The Software Engineering Body of Knowledge (SWEBOK) was published by the Institute of Electrical and Electronics Engineers (IEEE) [28] and also was published by the International Organization for Standardization (ISO) and International Electrotechnical Commission (IEC) as the international standard ISO/IEC TR 19759:2015 [32]. In this document, the body of knowledge in software engineering is defined.

The IEEE SWEBOK structures and defines the knowledge needed in software engineering in 15 knowledge areas (KA):

1. Software requirements
2. Software design
3. Software construction
4. Software testing
5. Software maintenance
6. Software configuration management
7. Software engineering management
8. Software engineering process
9. Software engineering models and methods
10. Software quality
11. Software engineering professional practice
12. Software engineering economics
13. Computing foundations
14. Mathematical foundations
15. Engineering foundations

The contents of each KA are described in depth one in each chapter of the IEEE SWEBOK. The structure of the chapters and, in consequence, in description of each KA is the following: First a table list of acronyms related to the KA is presented; then an introduction to the KA; then the wider part of the chapter composed be the breakdown of topics of the KA, with multiple subsections for each topic; then matrix with the relationships between the topics of the KA and the reference material; and finally a list of further readings and the references.

In addition, the IEEE SWEBOK, in the *Introduction to the Guide*, refers to seven additional disciplines that intersect with software engineering and about which KA descriptions in this guide may make reference to them.

These seven related disciplines are:

1. Computer engineering
2. Systems engineering
3. Project management
4. Quality management
5. General management

6. Computer science
7. Mathematics

Even if software engineers should have knowledge of the seven disciplines above, however, the guide does not include the knowledge of these disciplines because they are out of scope of software engineering.

SWEBOK has been used, for example, to develop data science engineering knowledge area group; the KAs used were software requirements, software design, software engineering process, software engineering models and methods, software quality, and specifically, in the data science (big data) applications design knowledge area, the SWEBOK selected KAs used were software requirements, software design, software construction, software testing, software maintenance, software configuration management, software engineering management, software engineering process, software engineering models and methods, software quality, agile development technologies, methods, platforms and tools, DevOps and continuous deployment and improvement paradigm.

3.1.4   Business Analytics Body of Knowledge (IIBA BABOK)

The Business Analysis Body of Knowledge (BABOK) [29] was published by International Institute of Business Analysis (IIBA). It is the globally recognised standard for the practice of business analysis. BABOK Guide reflects the collective knowledge of the business analysis community and presents the most widely accepted business analysis practices. In this document, the body of knowledge in business analytics is defined.

The IIBA BABOK structures and defines the knowledge needed in business analytics in 6 knowledge areas (KA):

1. Business analysis planning and monitoring
2. Elicitation and collaboration
3. Requirements Lifecycle management
4. Strategy analysis
5. Requirements analysis and design definition
6. Solution evaluation

The contents of each KA logically organise tasks. Each task describes the typical knowledge, skills, deliverables and techniques that the business analyst requires to be able to perform those tasks competently: KA1 describes the tasks used to organise and coordinate business analysis efforts; KA2 describes the tasks used to prepare for and conduct elicitation activities and confirm the results; KA3 describes the tasks used to manage and maintain requirements and design information from inception to retirement; KA4 describes the tasks used to identify the business need, address that need and align the change strategy within the enterprise; KA5 describes the tasks used to organise requirements, specify and model requirements and designs, validate and verify information, identify solution options and estimate the potential value that could be realised; and finally KA6 describes the tasks used to assess the performance of and value delivered by a solution and to recommend improvements on increasing value.

Although the KAs and their business analysis tasks are the core content of the IIBA BABOK, the guide not only includes the chapter of *knowledge areas* but includes four chapters more:

- *Business analysis key concepts.* In this chapter, important terms that are the foundation of the practice of business analysis are defined.
- *Underlying competencies.* In this chapter, the behaviours, characteristics, knowledge and personal qualities that help business analysts be effective in their job are described.
- *Techniques.* In this chapter, 50 of the most common techniques used by business analysts are described.
- *Perspectives.* In this chapter, five different views of business analysis are described: agile, business intelligence, information technology, business architecture and business process management.

BABOK provides interesting example of business-oriented body of knowledge that is important for data science knowledge domain. It has been used, for example, to develop the business analytics knowledge area group, and specifically in the business analytics foundation knowledge area; the KAs used were business analysis planning and monitoring, requirements analysis and design definition, requirements lifecycle management (from inception to retirement), solution evaluation and improvements recommendation.

3.1.5 Data Management Body of Knowledge (DAMA DMBOK)

The Data Management Body of Knowledge (DMBOK) [30] was published by the Data Management Association International (DAMAI). In this document, the body of knowledge in data management is defined, and also a Data Management Dictionary of Terms is provided.

The DAMA DMBOK structures and defines the knowledge needed in data management in 11 knowledge areas (KA):

1. Data governance
2. Data architecture
3. Data modelling and design
4. Data storage and operations
5. Data security
6. Data integration and interoperability
7. Documents and content

 8. Reference and master data
 9. Data warehousing and business intelligence
10. Metadata
11. Data quality

Each KA has section topics that logically group activities and is described by a context diagram. Each context diagram includes:

- Definition. A concise description of the KA.
- Goals. The desired outcomes of the KA within this topic.
- Process. The list of discrete activities and sub-activities to be performed, with activity group indicators.
- Inputs. Documents or raw materials are directly necessary for a process to initiate or continue.
- Supplier roles. Roles and/or teams that supply the inputs to the process.
- Responsible. Roles and/or teams that perform the process.
- Stakeholder. Roles and/or teams informed or consulted on the process execution.
- Tools. Technology types used by the process.
- Deliverables. What is directly produced by the processes.
- Consumer roles. Roles and/or teams that expect and receive the deliverables.
- Metrics. Measurements that quantify the success of processes based on the goals.

Although the KAs and their data management group activities are the core content of the DAMA DMBOK, the guide also includes an additional Data Management section containing topics that describe the knowledge requirements for data management professionals.

Among the proposed uses by DAMA of the BoK are *guiding efforts to implement and improve data management knowledge areas and guiding the development and delivery of data management curriculum content for higher education.*

DMBOK has been used, for example, to develop the data management knowledge area group, and specifically the data management and enterprise data infrastructure knowledge area, and the data governance knowledge area; the KAs used were data governance, data architecture, data modelling and design, data storage and operations, data security, data integration and interoperability, documents and content, reference and master data, data warehousing and business intelligence, metadata, data quality. When using DM-BoK it needs to be extended with the recent data modelling technologies and big data management platforms that address generic big data properties such as volume, veracity, velocity. New data security and privacy protections need to be addressed as well.

3.1.6 Project Management Professional Body of Knowledge (PMBOK)

The Project Management Professional Body of Knowledge (PMBOK) [31] was published by the Project Management Institute (PMI) and also was published by the American National Standards Institute (ANSI) as standard, and finally the International Organization for Standardization (ISO) as the international standard ISO/IEC 21500:2012 [33]. In this document, the body of knowledge in project management is defined.

The PMI PMBOK structures and defines the knowledge needed in project management in 5 process groups and 10 knowledge areas (KA). Each process group contains processes within some or all the knowledge areas. Each of the 42 processes has:

- Inputs
- Tools
- Techniques
- Outputs

The 5 process groups of the PMI PMBOK are:

1. Initiating
2. Planning
3. Executing
4. Monitoring and controlling
5. Closing

The ten knowledge areas (KA), linked to the process groups, of the PMI PMBOK are:

1. Project integration management
2. Project scope management
3. Project time management
4. Project cost management
5. Project quality management
6. Project human resource management
7. Project communications management
8. Project risk management
9. Project procurement management
10. Project stakeholder management

PMBOK has been used, for example, to develop the research methods and project management knowledge area group, and specifically the project management knowledge area; the KAs used have been PMI-BoK selected KAs, project integration management, project scope management, project quality, project risk management.

3.2   Definition of a Data Science Body of Knowledge (DS-BoK)

The DS-BoK contains the following knowledge area groups (KAGs) that follows the competence groups defined in Chap. 2:

1. DSDA, Data Analytics
2. DSENG, Data Engineering
3. DSDM, Data Management
4. DSRMP, Research methods and project management for research-related professions and business process management for business-related *professions*
5. DSBA, Business Analytics

The subject domain-related knowledge group (scientific or business) KAG*-DSBA is recognised as essential for practical work of data scientist which in fact means not professional work in a specific subject domain but understanding the domain-related concepts, models and organisation and corresponding data analysis methods and models. These knowledge areas will be a subject for future development in tight cooperation with subject domain specialists.

3.2.1   Data Analytics Knowledge Area Group (KAG1-DSDA)

The KAG1-DSDA data analytics knowledge area group is key and distinguishing KAG for DS-BoK. It includes different methods and algorithms, primarily statistical, machine learning and data mining, to enable data processing, modelling, analysis and inspection with the goal of discovering useful information, providing insight and recommendations and supporting decision-making. The following are the six commonly defined the data science analytics knowledge areas (KA):

1. KA01.01 (DSDA.01/SMA) Statistical methods for data analysis
2. KA01.02 (DSDA.02/ML) Machine learning
3. KA01.03 (DSDA.03/DM) Data mining
4. KA01.04 (DSDA.04/TDM) Text data mining
5. KA01.05 (DSDA.05/PA) Predictive analytics
6. KA01.06 (DSDA.06/MSO) Computational modelling, simulation and optimisation

3.2.1.1   Statistical Methods Knowledge Area

KA01.01 (DSDA.01/SMA) Statistical methods. This knowledge area includes descriptive statistics, exploratory data analysis (EDA) focused on discovering new features in the data and confirmatory data analysis (CDA) dealing with validating formulated hypotheses.

Starting with and initial KU about a general overview and main concepts, the 16 suggested specific knowledge units (KU) for the statistical method knowledge are:

1. KU1.01.00. General overview and main concepts in statistical methods for data analysis
2. KU1.01.01. Probability and statistics
3. KU1.01.02. Statistical paradigms (regression, time series, dimensionality, clusters)
4. KU1.01.03. Probabilistic representations (causal networks, Bayesian analysis, Markov nets)
5. KU1.01.04. Frequentist and Bayesian statistics
6. KU1.01.05. Probabilistic reasoning
7. KU1.01.06. Exploratory and confirmatory data analysis
8. KU1.01.07. Quantitative analytics
9. KU1.01.08. Qualitative analytics
10. KU1.01.09. Data preparation and preprocessing
11. KU1.01.10. Performance analysis
12. KU1.01.11. Markov models, Markov networks
13. KU1.01.12. Operations research
14. KU1.01.13. Information theory
15. KU1.01.14. Discrete mathematics and graph theory
16. KU1.01.15. Mathematical analysis
17. KU1.01.16. Mathematical software and tools

3.2.1.2 Machine Learning Methods Knowledge Area

KA01.02 (DSDA.02/ML) Machine learning knowledge area. Machine learning and related methods for information search, image recognition, decision support, classification.

Starting with and initial KU about a general overview and main concepts, the 13 suggested specific knowledge units (KU) for the machine learning methods knowledge are:

1. KU1.02.00. General overview and main concepts in machine learning
2. KU1.02.01. Machine learning theory and algorithms
3. KU1.02.02. Supervised machine learning
4. KU1.02.03. Unsupervised machine learning
5. KU1.02.04. Reinforced learning
6. KU1.02.05. Classification methods
7. KU1.02.06. Design and analysis of algorithms
8. KU1.02.07. Game theory and mechanism design
9. KU1.02.08. Artificial intelligence
10. KU1.01.02. Statistical paradigms (regression, time series, dimensionality, clusters)
11. KU1.01.03. Probabilistic representations (causal networks, Bayesian analysis, Markov nets)

12. KU1.01.04. Frequentist and Bayesian statistics
13. KU1.01.05. Probabilistic reasoning
14. KU1.01.08. Performance analysis

3.2.1.3   Data Mining Knowledge Area

KA01.03 (DSDA.03/DM) Data mining knowledge area. It is a particular data analysis technique that focuses on modelling and knowledge discovery for predictive rather than purely descriptive purposes.

Starting with and initial KU about a general overview and main concepts, the 13 suggested specific knowledge units (KU) for the data mining knowledge are:

1. KU1.03.00. General overview and main concepts in data mining
2. KU1.03.01. Data mining and knowledge discovery
3. KU1.03.02. Knowledge representation and reasoning
4. KU1.03.03. CRISP-DM and data mining stages
5. KU1.03.04. Anomaly detection
6. KU1.03.05. Time series analysis
7. KU1.03.06. Feature selection, a priori algorithm
8. KU1.03.07. Graph data analytics
9. KU1.01.08. Performance analysis
10. KU1.02.01. Machine learning theory and algorithms
11. KU1.02.02. Supervised machine learning
12. KU1.02.03. Unsupervised machine learning
13. KU1.02.04. Reinforced learning
14. KU1.02.05. Classification methods

3.2.1.4   Text Data Mining Knowledge Area

KA01.04 (DSDA.04/TDM) Text analytics applies statistical, linguistic and structural techniques to extract and classify information from textual sources, a species of unstructured data.

Starting with and initial KU about a general overview and main concepts, the seven suggested specific knowledge units (KU) for the text data mining knowledge are:

1. KU1.04.00. General overview and main concepts in text data mining
2. KU1.04.01. Text analytics including statistical, linguistic and structural techniques to analyse structured and unstructured data
3. KU1.04.02. Data mining and text analytics
4. KU1.04.03. Natural language processing
5. KU1.04.04. Predictive models for text
6. KU1.04.05. Retrieval and clustering of documents
7. KU1.04.06. Information extraction
8. KU1.04.07. Sentiments analysis

3.2.1.5   Predictive Analytics Knowledge Area

KA01.05 (DSDA.05/PA) Predictive analytics knowledge area. It focuses on application of statistical models for predictive forecasting or classification.

Starting with and initial KU about a general overview and main concepts, the seven suggested specific knowledge units (KU) for the predictive analytics knowledge are:

1. KU1.05.00. General overview and main concepts in predictive analytics
2. KU1.05.01. Predictive modelling and analytics
3. KU1.05.02. Inferential and predictive statistics
4. KU1.05.03. Machine learning for predictive analytics
5. KU1.05.04. Regression and multi analysis
6. KU1.05.05. Generalised linear models
7. KU1.05.06. Time series analysis and forecasting
8. KU1.05.07. Deploying and refining predictive models

3.2.1.6   Computational Modelling, Simulation and Optimisation Knowledge Area

KA01.06 (DSDA.06/MSO) Computational modelling, simulation and optimisation knowledge area.

Starting with and initial KU about a general overview and main concepts, the five suggested specific knowledge units (KU) for the business analytics and business intelligence knowledge are:

1. KU1.06.00. General overview and main concepts in computational modelling, simulation and optimisation
2. KU1.06.01. Modelling and simulation theory and techniques (general and domain oriented)
3. KU1.06.02. Operations research and optimisation
4. KU1.06.03. Large-scale modelling and simulation systems
5. KU1.06.04. Network optimisation
6. KU1.06.05. Risk simulation and queueing

3.2.2   Data Engineering Knowledge Area Group (KAG2-DSENG)

The KAG2-DSENG group includes selected KAs from ACM CS-BoK and SWEBOK and extends them with new technologies and engineering technologies and paradigm such as cloud based, agile technologies and DevOps that are promoted as continuous deployment and improvement paradigm and allow organisations to implement agile business and operational models. The following are the seven commonly defined the data science engineering knowledge areas:

1. KA02.01 (DSENG.01/BDIT) Big data infrastructure and technologies
2. KA02.02 (DSENG.02/DSIAPP) Infrastructure and platforms for data science applications
3. KA02.03 (DSENG.03/CCT) Cloud computing technologies for big data and data analytics
4. KA02.04 (DSENG.04/SEC) Data and applications security
5. KA02.05 (DSENG.05/BDSE) Big data systems organisation and engineering
6. KA02.06 (DSENG.06/DSAPPD) Data science (big data) applications design
7. KA02.07 (DSENG.07/IS) Information systems (to support data-driven decision-making)

3.2.2.1   Big Data Infrastructure and Technologies Knowledge Area

Starting with and initial KU about a general overview and main concepts, the 10 suggested specific knowledge units (KU) for the big data infrastructure and technologies knowledge are:

1. KU2.01.00. General overview and main concepts in big data infrastructure and technologies
2. KU2.01.01. Computer systems organisation for big data applications, CAP, BASE and ACID theorems
3. KU2.01.02. Parallel and distributed computer architecture
4. KU2.01.03. High performance and cloud computing
5. KU2.01.04. Clouds and scalable computing
6. KU2.01.05. Cloud-based big data platforms and services
7. KU2.01.06. Big data (large-scale) storage and file systems (HDFS, Ceph, etc.)
8. KU2.01.07. NoSQL databases
9. KU2.01.08. Computer networks for high-performance computing and big data infrastructure
10. KU2.01.09. Computer networks: architectures and protocols
11. KU2.01.10. Big data infrastructure management and operation

3.2.2.2   Infrastructure and Platforms for Data Science Applications Knowledge Area

Starting with and initial KU about a general overview and main concepts, the eight suggested specific knowledge units (KU) for the infrastructure and platforms for data science applications knowledge are:

1. KU2.02.00. General overview of infrastructure and platforms for data science applications
2. KU2.02.01. Big data infrastructure: services and components, including data storage infrastructure

3. KU2.02.02. Big data analytics platforms and tools (including Hadoop, Spark and cloud-based big data services)
4. KU2.02.03. Large-scale cloud-based storage and data management
5. KU2.02.04. Cloud-based applications and services operation and management
6. KU2.02.05. Big data and cloud-based systems design and development
7. KU2.02.06. Data processing models (batch, steaming, parallel)
8. KU2.02.07. Enterprise information systems
9. KU2.02.08. Data security and protection

3.2.2.3   Cloud Computing Technologies for Big Data and Data Analytics Knowledge Area

Starting with and initial KU about a general overview and main concepts, the three suggested specific knowledge units (KU) for the cloud computing technologies for big data and data analytics knowledge are:

1. KU2.03.00. General overview of cloud computing technologies and their use for big data and data analytics
2. KU2.03.01. Cloud computing architecture and services
3. KU2.03.02. Cloud computing engineering (infrastructure and services design, management and operation)
4. KU2.03.03. Cloud-based applications and services operation and management

3.2.2.4   Data and Applications Security Knowledge Area

Starting with and initial KU about a general overview and main concepts, the six suggested specific knowledge units (KU) for the data and applications security knowledge are:

1. KU2.04.00. General overview and main concepts in data and applications security
2. KU2.04.01. Infrastructure, applications and data security
3. KU2.04.02. Data encryption and key management, blockchain-based technologies
4. KU2.04.03. Access control and identity management
5. KU2.04.04. Security services management, including compliance and certification
6. KU2.04.05. Data anonymisation
7. KU2.04.06. Data privacy

3.2.2.5   Big Data Systems Organisation and Engineering Knowledge Area

Starting with and initial KU about a general overview and main concepts, the 11 suggested specific knowledge units (KU) for the big data systems organisation and engineering knowledge are:

1. KU2.05.00. General overview and main principles of big data systems organisations and engineering
2. KU2.05.01. Big data systems organisation and design
3. KU2.05.02. Big data algorithms for large-scale data processing
4. KU2.05.03. Big data analytics
5. KU2.05.04. Big data analytics platforms and tools (including Hadoop, Spark and cloud-based big data services)
6. KU2.05.05. Big data algorithms for data ingest, preprocessing and visualisation
7. KU2.05.06. Big data systems for application domains
8. KU2.05.07. Big data software (systems) architectures
9. KU2.05.08. Requirements engineering and software systems development
10. KU2.05.09. Large and ultra-large-scale software systems organisation
11. KU2.05.10. DevOps and cloud-enabled applications development
12. KU2.05.11. Big data infrastructure management and operation

3.2.2.6   Data Science (Big Data) Applications Design Knowledge Area

Starting with and initial KU about a general overview and main concepts, the eight suggested specific knowledge units (KU) for the data science (big data) applications design knowledge are:

1. KU2.06.00. General overview and main approaches to data science (big data) applications design
2. KU2.06.01. Data analytics, data handling software requirements and design
3. KU2.06.02. Applications engineering management
4. KU2.06.03. Software engineering models and methods
5. KU2.06.04. Software quality assurance
6. KU2.06.05. Programming languages for big data analytics: R, python, Pig, Hive, others
7. KU2.06.06. Models and languages for complex interlinked data presentation and visualisation
8. KU2.06.07. Agile development methods, platforms and tools
9. KU2.06.08. DevOps and continuous deployment and improvement paradigm

3.2.2.7 Information Systems (To Support Data-Driven Decision Making) Knowledge Area

Starting with and initial KU about a general overview and main concepts, the eight suggested specific knowledge units (KU) for the information systems (to support data-driven decision making) knowledge are:

1. KU2.07.00. General overview and basic architectures of information systems to support data-driven decisions)
2. KU2.07.01. Decision analysis and decision support systems
3. KU2.07.02. Predictive analytics and predictive forecasting
4. KU2.07.03. Data analysis and statistics
5. KU2.07.04. Data warehousing and data mining
6. KU2.07.05. Data mining
7. KU2.07.06. Multimedia information systems
8. KU2.07.07. Enterprise information systems
9. KU2.07.08. Collaborative and social computing systems and tools

3.2.3 Data Management Knowledge Area Group, KAG3-DSDM

The KAG3-DSDM group includes most of KAs from DM-BoK, however, it is extended with KAs related to RDA recommendations, community data management models (open access, open data, etc.) and general data lifecycle management that is used as a central concept in many data management-related education and training courses. The following are the six commonly defined the data management knowledge areas:

1. KA03.01 (DSDM.01/DMORG) General principles and concepts in data management and organisation
2. KA03.02 (DSDM.02/DMS) Data management systems
3. KA03.03 (DSDM.03/EDMI) Data management and enterprise data infrastructure
4. KA03.04 (DSDM.04/DGOV) Data governance
5. KA03.05 (DSDM.05/BDST0R) Big data storage (large scale)
6. KA03.06 (DSDM.05/DLIB) Digital libraries and archives

3.2.3.1 General Principles and Concepts in Data Management and Organisation Knowledge Area

Starting with and initial KU about a general overview and main concepts, the seven suggested specific knowledge units (KU) for the general principles and concepts in data management and organisation knowledge are:

1. KU3.01.00. Overview of general principles, concepts and practices in data management and organisation
2. KU3.01.01. Data type registries, PID, metadata
3. KU3.01.02. Data lifecycle management
4. KU3.01.03. Data infrastructure and data factories
5. KU3.01.04. Research data infrastructure, open science, open data, open access, ORCID
6. KU3.01.05. Data infrastructure compliance and certification
7. KU3.01.06. Ethical principle and data privacy
8. KU3.01.07. FAIR (findable, accessible, interoperable) principles in data management

3.2.3.2   Data Management Systems Knowledge Area

Starting with and initial KU about a general overview and main concepts, the eight suggested specific knowledge units (KU) for the data management systems knowledge are:

1. KU3.02.00. General overview and main architectural components in data management systems
2. KU3.02.01. Data architectures (OLAP, OLTP, ETL)
3. KU3.02.02. Data modelling, databases and database management systems
4. KU3.02.03. Data structures
5. KU3.02.04. Data models and query languages
6. KU3.02.05. Database design and models
7. KU3.02.06. Database administration
8. KU3.02.07. Data warehouses
9. KU3.02.08. Middleware for databases

3.2.3.3   Data Management and Enterprise Data Infrastructure Knowledge Area

Starting with and initial KU about a general overview and main concepts, the 10 suggested specific knowledge units (KU) for the data management and enterprise data infrastructure knowledge are:

1. KU3.03.00. General overview and main components in enterprise infrastructure for data management
2. KU3.03.01. Data management, including reference and master data
3. KU3.03.02. Data warehousing and business intelligence
4. KU3.03.03. Data storage and operations
5. KU3.03.04. Data archives/storage compliance and certification
6. KU3.03.05. Metadata, linked data, provenance
7. KU3.03.06. Data infrastructure, data registries and data factories
8. KU3.03.07. Data security and protection

9. KU3.03.08. Data backup
10. KU3.03.09. Data anonymisation
11. KU3.03.10. Data privacy

3.2.3.4 Data Governance Knowledge Area

Starting with and initial KU about a general overview and main concepts, the seven suggested specific knowledge units (KU) for the data management and enterprise data infrastructure knowledge are:

1. KU3.04.00. General overview and main concepts in data governance
2. KU3.04.01. Data governance, data quality, data Integration and interoperability
3. KU3.04.02. Data management planning
4. KU3.04.03. Data management policy
5. KU3.04.04. Data interoperability
6. KU3.04.05. Data curation
7. KU3.04.06. Data provenance
8. KU3.04.07. Responsible data use, data privacy, ethical principles, IPR, legal issues

3.2.3.5 Big Data Storage (Large-Scale) Knowledge Area

Starting with and initial KU about a general overview and main concepts, the five suggested specific knowledge units (KU) for the big data storage (large-scale) knowledge are:

1. KU3.05.00. General overview and architecture components in big data storage
2. KU3.05.01. Big data storage infrastructure and operations
3. KU3.05.02. Storage architectures, distributed files systems (HDFS, Ceph, Lustre, Gluster, etc.)
4. KU3.05.03. Data storage redundancy and backup
5. KU3.05.04. Data factories, data pipelines
6. KU3.05.05. Cloud-based storage, data lakes

3.2.3.6 Data Libraries, Data Archives, Digital Libraries Knowledge Area

Starting with and initial KU about a general overview and main concepts, the five suggested specific knowledge units (KU) for the big data storage (large-scale) knowledge are:

1. KU3.06.00. General overview of data libraries, data archives, digital libraries
2. KU3.06.01. Data libraries and data archives organisation and services
3. KU3.06.02. Digital libraries organisation and services
4. KU3.06.03. Information retrieval

5. KU3.06.04. Data curation and provenance
6. KU3.06.05. Search engines and technologies

3.2.4   Research Methods and Project Management Knowledge Area Group (KAG4-DSRMP)

The following are the two commonly defined the data management knowledge areas of the KAG4-DSRMP group:

1. KA04.01 (DSRMP.01/RM) Research methods
2. KA04.02 (DSRMP.02/PM) Project management

3.2.4.1   Research Methods Knowledge Area

Starting with and initial KU about a general overview and main concepts, the five suggested specific knowledge units (KU) for the research methods knowledge are:

1. KU4.01.00. Overview research methods and data-driven research
2. KU4.01.01. Research methodology, paradigms and research cycle
3. KU4.01.02. Modelling and experiment planning
4. KU4.01.03. Data selection and quality evaluation
5. KU4.01.04. Data life cycle
6. KU4.01.05. Use case analysis: research infrastructures and projects

3.2.4.2   Project Management Knowledge Area

Starting with and initial KU about a general overview and main concepts, the four suggested specific knowledge units (KU) for the project management knowledge are:

1. KU4.02.00. Overview research process and project management
2. KU4.02.01. Project integration management
3. KU4.02.02. Project scope management
4. KU4.02.03. Project quality
5. KU4.02.04. Project risk management

3.2.5   Business Analytics Knowledge Area Group (KAG5-DSBA)

The following are the two commonly defined the business analytics knowledge areas of the KAG5-DSBA group:

1. KA05.01 (DSBA.01/BAF) Business analytics foundation
2. KA05.02 (DSBA.02/BAEM) Business analytics organisation and enterprise management

3.2.5.1   Business Analytics Foundation Knowledge

Starting with and initial KU about a general overview and main concepts, the nine suggested specific knowledge units (KU) for the business analytics foundation knowledge are:

1. KU5.01.00. Overview business analytics methods and practices
2. KU5.01.01. Business analytics and business intelligence: data, models (statistical) and decisions
3. KU5.01.02. Data-driven customer relations management (CRP), user experience (UX) requirements and design
4. KU5.01.03. Operations analytics
5. KU5.01.04. Business process optimisation
6. KU5.01.05. Data warehouses technologies, data integration and analytics
7. KU5.01.06. Data-driven marketing technologies
8. KU5.01.07. Business Analytics Capstone
9. KU5.01.08. Econometrics methods and application for business analytics
10. KU5.01.09. Cognitive technologies for business analytics

3.2.5.2   Business Analytics Organisation and Enterprise Management Knowledge Area

Starting with and initial KU about a general overview and main concepts, the eight suggested specific knowledge units (KU) for the business analytics organisation and enterprise management knowledge are:

1. KU5.02.00. Overview of business analytics process organisation and enterprise management
2. KU5.02.01. Business processes and operations
3. KU5.02.02. Project scope and risk management
4. KU5.02.03. Business analysis planning and monitoring
5. KU5.02.04. Requirements analysis and design definition
6. KU5.02.05. Requirements lifecycle management (from inception to retirement)
7. KU5.02.06. Solution evaluation and improvements recommendation
8. KU5.02.07. Agile data-driven methodologies, processes and enterprises
9. KU5.02.08. Use case analysis: business and industry

3.3   Use of ACM CCS2012 in the DS-BoK and Proposed Extension of ACM CCS2012 from the DS-BoK

This section provides historical information about subset of the ACM Computing Classification System (ACM CCS2012) [14] taxonomy that provided the initial structure for the DS-BoK that was further extended with full set of knowledge areas and knowledge units related to data science that can be partly mapped to ACM CCS2012. The defined below subset of ACM CCS2012 classification can provide the basis for future ACM CCS2012 extension with a new classification group related to data science and individual disciplines that are missing in the current ACM-IEEE classification.

The ACM CCS2012 has been developed as a poly-hierarchical ontology that can be utilised in semantic web applications. It replaces the traditional 1998 version of the ACM CCS, which has served as the de facto standard classification system for the computing field for many years (also been more human readable). The ACM CCS2012 is being integrated into the search capabilities and visual topic displays of the ACM Digital Library. It relies on a semantic vocabulary as the single source of categories and concepts that reflect the state of the art of the computing discipline and is receptive to structural change as it evolves in the future. ACM provides a tool within the visual display format to facilitate the application of 2012 CCS categories to forthcoming papers and a process to ensure that the CCS stays current and relevant.

However, at the moment none of data science, big data or data-intensive science technologies are reflected in the ACM classification. The following is an extraction of possible classification facets from ACM CCS2012 related to data science what reflects multi-subject areas nature of data science:

As an example, cloud computing that is also a new technology and closely related to big data technologies is currently classified in ACM CCS2012 into 3 groups:

- Networks:: Network services:: Cloud computing
- Computer systems organisation:: Architectures:: Distributed architectures:: Cloud computing
- Software and its engineering:: Software organisation and properties:: Software systems structures:: Distributed systems organising principles:: Cloud computing

Taxonomy is required to consistently present information about scientific disciplines and knowledge areas related to data science. Taxonomy is an important component to link such components as data science competences and knowledge areas, body of knowledge and corresponding academic disciplines. From a practical point of view, taxonomy includes vocabulary of names (or keywords) and hierarchy of their relations.

The presented ACM CCS2012 subsets/subtrees contain scientific disciplines related to three data science knowledge area groups as they are defined in DS-BoK:

- KAG1-DSDA: Data analytics group including machine learning, statistical methods and business analytics
- KAG2-DSENG: Data science engineering group including software engineering and infrastructure engineering
- KAG3-DSDM: Data management group including data curation, preservation and data infrastructure

Two other groups KAG4-DSRMP, research methods and project management and KAG5-DSDK, do not have direct mapping to ACM CCS2012, and their taxonomies are defined based on other domain-specific bodies of knowledge. It is important to notice that ACM CCS2012 provides a top-level classification entry *applied computing* that can be used as an extension point for domain-related knowledge area group KAG5-DSDK.

The following approach was used when constructing the proposed taxonomy:

- ACM CCS2012 provides almost full coverage of data science-related knowledge areas or disciplines related to KAG1, KAG2 and KAG3. The following top-level classification groups are used:

  - Theory of computation
  - Mathematics of computing
  - Computing methodologies
  - Information systems
  - Computer systems organisation
  - Software and its engineering

- Each of KAGs includes subsets from few ACM CCS2012 classification groups to cover theoretical, technology, engineering and technical management aspects.
- Extension points are suggested for possible future extensions of related KAGs together with their hierarchies.
- KAG3-DSDM: Data management group is extended with new concepts and technologies developed by Research Data Alliance community and documented in community best practices.

In the following lists, the ACM CCS2012 classification facets related to the data science grouped by DS-BoK knowledge area groups and knowledge areas are presented.

3.3.1 Data Analytics

3.3.1.1 ACM CCS2012 Subjects Used to Develop the DS-BoK Data Analytics

Data science analytics-related scientific subjects from CCS2012 are:

- CCS2012: Computing methodologies
- CCS2012: Mathematics of computing

Statistical Methods Knowledge Area
Data science statistical methods related scientific subjects from CCS2012 are:

- Mathematics of computing
  - Discrete mathematics

    Graph theory
    Probability and statistics
    Probabilistic representations
    Probabilistic inference problems
    Probabilistic reasoning algorithms
    Probabilistic algorithms

  - Statistical paradigms
  - Mathematical software
  - Information theory
  - Mathematical analysis

Machine Learning Methods Knowledge Area
Data science machine learning methods related scientific subjects from CCS2012 are:

For the KU1.02.00 to KU1.02.08:

- Computing methodologies
  - Artificial intelligence

    Machine learning
    Learning paradigms

  - Supervised learning
  - Unsupervised learning
  - Reinforcement learning
  - Multi-task learning
  - Machine learning approaches

    Machine learning algorithms

For KU1.01.02, KU1.01.03, KU1.01.04, KU1.01.05 and KU1.01.08:

- Theory of computation
  - Design and analysis of algorithms

    Data structures design and analysis

  - Theory and algorithms for application domains

    Machine learning theory
    Algorithmic game theory and mechanism design

  - Semantics and reasoning

Data Mining Knowledge Area

Data Science data mining related scientific subjects from CCS2012 are:

- Theory of computation
  - Design and analysis of algorithms

    Data structures design and analysis

  - Theory and algorithms for application domains

    Machine learning theory
    Algorithmic game theory and mechanism design

  - Semantics and reasoning

Text Data Mining Knowledge Area

Data science text mining related scientific subjects from CCS2012 are:

- Computing methodologies
  - Artificial intelligence

    Natural language processing
    Knowledge representation and reasoning
    Search methodologies

Predictive Analytics Knowledge Area

Data science predictive analytics related scientific subjects from CCS2012 are:

- Computing methodologies
  - Artificial intelligence

    Natural language processing
    Knowledge representation and reasoning
    Search methodologies

Computational Modelling, Simulation and Optimisation Knowledge Area

Data science computational modelling, simulation and optimisation related scientific subjects from CCS2012 are:

- Computing methodologies
  - Modelling and simulation
  - Model development and analysis
  - Simulation theory
  - Simulation types and techniques
  - Simulation support systems

3.3.1.2 ACM CCS2012 Extension Points from the DS-BoK Data Analytics

Theory of Computation
ACM CCs 2012 Theory of computation extension point from DS-Bok is:

- Algorithms for big data computation

Mathematics of Computing
ACM CCs 2012 Mathematics of computing extension point from DS-Bok is:

- Mathematical software for big data computation

Computing Methodologies
ACM CCs 2012 Computing methodologies extension point from DS-Bok is:

- New DSA computing

Information Systems
ACM CCs 2012 Information systems extension point from DS-Bok is:

- Big data systems (e.g. cloud based)

    ACM CCs 2012 Information systems applications extension point from DS-Bok are:

- Big data applications
- Doman-specific data applications

3.3.2 Data Engineering

3.3.2.1 ACM CCS2012 Subjects Used to Develop the DS-BoK Data Engineering

Data science engineering related scientific subjects from CCS2012 are:

- CCS2012: Computer systems organisation
- CCS2012: Information systems
- CCS2012: Software and its engineering

Big Data Infrastructure and Technologies Knowledge Area
Data science big data infrastructure and technologies related scientific subjects from CCS2012 are:

- Computer systems organisation
- Architectures

- – Parallel architectures
- – Distributed architectures

- Networks

  - – Network architectures
  - – Network services
  - – Cloud computing

Infrastructure and Platforms for Data Science Applications Knowledge Area

Data science infrastructure and platforms for data science applications related scientific subjects from CCS2012 are:

For the KU2.02.07:

- Information systems
- Information storage systems
- Information systems applications

Big Data Systems Organisation and Engineering Knowledge Area

Data science big data systems organisation and engineering related scientific subjects from CCS2012 are:

For the KU2.05.07 to KU2.05.11:

- Software and its engineering

  - – Software organisation and properties

      Software system structures

  - – Software architectures

      Software system models
      Distributed systems organising principles

  - – Cloud computing
  - – Grid computing
  - – Software notations and tools

      General programming languages
      Software creation and management

Information Systems (To Support Data-Driven Decision-Making) Knowledge Area

Data science information systems related scientific subjects from CCS2012 are:

- Information systems

  - – Information systems applications
  - – Decision support systems
  - – Data warehouses

- Expert systems
- Data analytics
- Online analytical processing
- Multimedia information systems
- Data mining

3.3.2.2   ACM CCS2012 Extension Points from the DS-BoK Data Engineering

From data engineering are proposed all the extension points of the ACM CCs 2012 proposed in data engineering and the following ones:

Software and Its Engineering
ACM CCs 2012 Software and its engineering extension point software organisation and properties from DS-Bok are:

- Big data applications design
- Data analytics programming languages

3.3.3   Data Management

3.3.3.1   ACM CCS2012 Subjects Used to Develop the DS-BoK Data Management

Data science management related scientific subjects from CCS2012 are:

- CCS2012: Information systems

Data Management Systems Knowledge Area
Data science data management systems related scientific subjects from CCS2012 are:

- Information systems

  - Data management systems

        Database design and models
        Data structures
        Database management system engines
        Query languages
        Database administration
        Middleware for databases
        Information integration

- Theory of computation

  - Database theory

Data Libraries, Data Archives, Digital Libraries Knowledge Area
Data science data libraries, data archives, digital libraries related scientific subjects from CCS2012 are:

- Information systems

  - Information systems applications
  - Digital libraries and archives

- Information retrieval

  - Document representation
  - Retrieval models and ranking
  - Search engine architectures and scalability
  - Specialised information retrieval

3.3.3.2 ACM CCS2012 Extension Points from the DS-BoK Data Management

From data management are proposed all the extension points of the ACM CCs 2012 proposed in data management and the following ones:

Information Systems
ACM CCs 2012 Information systems extension point from DS-Bok are:

- Data management systems

  - Data types and structures description
  - Metadata standards
  - Persistent identifiers (PID)
  - Data types registries

3.3.4 Research Methods and Project Management

There are not data science research methods and project management related scientific subjects from CCS2012, and, in consequence, no extensions have been proposed.

3.3.5 Domain Knowledge

3.3.5.1 ACM CCS2012 Subjects Used to Develop the DS-BoK Domain Knowledge

There are not domain knowledge related scientific subjects from CCS2012.

3.3.5.2 ACM CCS2012 Extension Points from the DS-BoK Domain Knowledge

From domain knowledge are proposed all the extension points of the ACM CCs 2012 proposed in data management and the following ones:

Applied Computing
ACM CCs 2012 Applied computing extension points from DS-Bok are:

- Physical sciences and engineering
- Life and medical sciences
- Law, social and behavioural sciences
- Computer forensics
- Arts and humanities
- Computers in other domains

Chapter 4
Data Science Curriculum

Tomasz Wiktorski, Yuri Demchenko, and Juan J. Cuadrado-Gallego

This section provides background information and best practices in building effective professional curricula for specific domains of knowledge, target groups and purposes. The reviewed selected learning model and curricula design models are used to develop the EDISON approach that is targeted to provide quality education and training for specific groups of data science-related professions to acquire necessary competences and skills.

This chapter has three sections with the following structure: Sect. 4.1 presents an overview of best practices in curricula design; Sect. 4.2 presents the definition of a Data Science Model Curriculum approach, with five subsections, one for each one if the five knowledge area groups defined for the DS-BoK; and finally Sect. 4.3 presents a developed Data Science Model Curriculum.

4.1 Overview of Best Practices in Curricula Design

The data science competences defined in Chap. 2 and the Data Science Body of Knowledge (DS-Bok), defined in Chap. 3, have been the main documents used to define the proposed approach to definition of the Model Curriculum in Data Science (MC-DS) structure and content. Curricula and bodies of knowledge have been

T. Wiktorski
Universitetet i Stavanger, Stavanger, Norway
e-mail: tomasz.wiktorski@uis.no

Y. Demchenko
Universiteit van Amsterdam, Amsterdam, The Netherlands
e-mail: y.demchenko@uva.nl

J. J. Cuadrado-Gallego (✉)
Department of Computer Science, University of Alcalá, Madrid, Spain
e-mail: jjcg@uah.es

© Springer Nature Switzerland AG 2020
J. J. Cuadrado-Gallego, Y. Demchenko (eds.), *The Data Science Framework*,
https://doi.org/10.1007/978-3-030-51023-7_4

reviewed to identify best practices and components to be used for the initial definition:

- ACM-IEEE Computer Science Curriculum and Body of Knowledge, ACM-IEEE CS2013 and CS-BoK) [25]
- Information Technology Competency Model of Learning Outcome ACM CCECC2014 [34]
- ICT professional Body of Knowledge and ICT leadership curriculum, EU ICT-BoK [27]

Also, other relevant bodies of knowledge (BoKs), which were also used in defining the DS-BoK defined in Chap. 3, have also been used to define the MS-DS:

- Software Engineering Body of Knowledge (IEEE SWEBOK) [28]
- Business Analytics Body of Knowledge (IIBA BABOK) [29]
- Data Management Body of Knowledge (DM-BoK) by Data Management Association International (DAMAI) [30]
- Project Management Professional Body of Knowledge (PMI PM-BoK) [31]

It is important to mention that due to the complex nature of the data science profession consisting of few quite different knowledge areas, the MC-DS definition requires combination of the elements from different BoKs and using different approaches to the curriculum definition; moreover, the MC-DS should allow different learning models and adaptation to different subject domains. The final curriculum definition will depend on local conditions defined by the job market demand side (i.e. employers, industry), available teaching staff and expertise and available educational base and infrastructure.

To define consistently the MC-DS, we need to understand the commonly accepted approaches to defining education and training programmes and put them in the context of the European education system and policies and also consider alignment with the international practices. Two approaches to education and training are followed in practice, the traditional approach which is based on defining the time students must spend learning a given topic or concept like the European Credit Transfer and Accumulation System (ECTS) [35], or Carnegie unit credit hour [36]. The former is also known as competence-based education or outcomes-based education (OBE), it is focusing on the outcome assessing whether students have mastered the given competences, namely the skills, abilities and knowledge.

There is no specified style of teaching or assessment in OBE; instead classes, opportunities and assessments should all help students achieve the specified outcomes. In 2012, the EC has called for a rethinking of education towards OBE approach. The motivation for such a rethinking is to ensure that education is more relevant to the needs of students and the labour market; assessment methods need to be adapted and modernised. Not like the traditional BoK which is defined in terms of knowledge areas (KA) in OBE, the BoK and curriculum are defined in terms of the core learning outcomes which are grouped into technical competence areas and workplace skills.

4.1.1   ACM-IEEE Computer Science Curricula (CS2013)

In the ACM-IEEE Computer Science Curricula (CS2013) has been treated in the previous Chap. 3, in Sect. 3.1.1. The document describes and structures the 18 knowledge areas needed to define a curriculum in computer science.

Jointly to the previous document, the ACM Curriculum for computing Education in Community Colleges [34] defines a BoK for IT outcome-based learning/education which identifies 6 technical competency areas and 5 workplace skills, defined in Chap. 2 "Competences". While the technical areas are specific to IT competences and specify a set of demonstrable abilities of graduates to perform some specific functions, the so-called workplace skills describe the ability of the student/trainee to:

- Function effectively as a member of a diverse team
- Read and interpret technical information
- Engage in continuous learning
- Professional, legal and ethical behaviour
- Demonstrate business awareness and workplace effectiveness

The ACM steering committee agrees on set principles to guide the development of CS2013 model curriculum. These principles aim at providing students with necessary flexibility to work across disciplines and prepare the graduates for a variety of disciplines. Following is the summary of the most important principles:

- CS2013 should provide guidance for the expected level of mastery of topics by the graduate.
- CS2013 should provide realistic, adoptable recommendations that provide guidance and flexibility allowing curricula designs that are innovative and track recent developments in the field.
- Size of the essential knowledge must be manageable.
- Computer science curricula should prepare graduates to succeed in a rapid changing area.
- CS2013 should identify the fundamental skills and knowledge that all computer science graduate should possess while providing the greatest flexibility in selecting topics.
- CS2013 should provide a great flexibility in organising topics into courses and curricula.

Through these principles ACM provides graduate with fundamental knowledge in the areas described in the ACM-IEEE BoK and a style of thinking and problem-solving. The latter is achieved through defining the expected characteristics of computer science graduate, namely:

- Technical understanding of computer science
- Familiarity with common themes and principles
- Appreciation of interplay between theory and practice
- System-level perspective
- Problem-solving skills

- Project experience
- Commitment to lifelong learning
- Commitment to professional responsibility
- Communication and organisation skills
- Appreciation of domain-specific knowledge

ACM-IEEE follows a simple straightforward approach to design the ACM Model Curriculum. It starts from the CS2013-based CS-BoK which is structured into knowledge areas (KA) organised in topical themes rather than by courses boundary. Each KA is further organised into a set of knowledge units (KU). In the final step, each KU lists a set of topics and learning outcomes (LO). The LO are associated with a level of mastery derived from the Bloom's taxonomy (familiarity, usage and assessment).

4.1.2 Information Technology Competency Model of Learning Outcome (ACM CCECC2014)

The ACM Committee for Computing Education in Community Colleges (CCECC) and its partner professional societies (in particular, IEEE Computer Society) have jointly produced curricular recommendations and guidelines for baccalaureate computing programmes, known collectively as the ACM Computing Curricula series. One of these guidelines is the Curriculum Guidelines for Undergraduate Degree Programs in Information Technology (IT2008) and its later published companion document ACM Competency Model of Core Learning Outcomes and Assessment for Associate-Degree Curriculum in Information Technology (IT2014) [37].

The guidelines use the competence-based learning model that focuses on the extent that students learn given competencies (knowledge, skills, qualifications), instead of focusing on so-called *seat time*, commonly expressed by credit points. The proposed competency model for constructing information technology curricula is based on defining measurable learning outcomes. The CCECC identified the body of knowledge as a set of fifty student learning outcomes that span the first three levels of Bloom's revised taxonomy (see above), and each outcome is accompanied by a three-tier assessment rubric that provides additional clarity and a measurable evaluation metric [37].

4.1.3 Foundational Information and Communication Technologies Body of Knowledge, EC ICT-BoK

The ICT-BoK [27] also has been treated in Chap. 3, Sect. 3.1.2, and it is an effort promoted by the European Commission, under the eSkills initiative (http://eskills4jobs.ec.europa.eu/) to define and organise the core knowledge of the ICT discipline. In order to foster the growth of digital jobs in Europe and to improve ICT professionalism, a study has been conducted to provide the basis of a *framework for ICT professionalism* (http://ictprof.eu/). This framework consists of four building blocks (also called pillars) which are also found in other professions:

1. Body of knowledge (BoK)
2. Competence framework
3. Education and training resources
4. Code of professional ethics

A competence framework already exists and consists in the e-Competence Framework (now in its version 3.0 and promoted by CEN). However, an ICT body of knowledge that provides the basis for a common understanding of the foundational knowledge an ICT professional should possess is not yet available.

The ICT-BoK aims at informing about the level of knowledge required to enter the ICT profession and acts as the first point of reference for anyone interested in working in ICT. Even if the ICT-BoK does not refer to data science competences explicitly, the identified ICT processes can be applied to data management processes both in industry and academia in the context of well-defined and structured projects.

Further ICT BoK development was focused on developing the new curricula for e-leadership skills in Europe (refer to the SCALE Project report [38] for details).

4.2   Definition of a Data Science Model Curriculum Approach

This section presents the definition of the EDISON Data Science Model Curriculum that is primarily based on mapping between DS-BoK knowledge areas and MC-DS learning units, which may represent academic courses and training modules, for required competence groups using competence-based learning model.

The proposed MC-DS can be used for defining individual curricula for specific data science professional profiles or customised individual curricula for practitioners that want to obtain a data science qualification or certification. The example of applying competence-based approach to selecting a set of learning units for different DSP profiles is given in Chap. 6. The proposed methods can be used for developing tools for customising or profiling the training and/or education programmes for students or individual trainees.

To build the Data Science Model Curriculum, the following theoretical foundations had been used: Bloom's taxonomy, constructive alignment and problem-based learning and competence-based learning model. We describe each of them shortly.

Bloom's taxonomy. Bloom's taxonomy [39] provides a conceptual framework to organise levels of learning of a topic or subject and assigns action verbs to each level that help to understand activities related to particular level of learning. The levels of learning are (below are shown typical attributes of the different levels of learning and example questions to test these levels):

- *Knowledge*: Exhibit memory of previously learned materials by recalling facts, terms, basic concepts and answers. Knowledge of specifics—terminology, specific facts. Knowledge of ways and means of dealing with specifics—conventions, trends and sequences, classifications and categories, criteria, methodology. Knowledge of the universals and abstractions in a field—principles and generalisations, theories and structures. Questions like: What are the main benefits of implementing big data and data analytics methods for organisation? What are main methods and algorithms in the supervised machine learning.
- *Comprehension*: Demonstrate understanding of facts and ideas by organising, comparing, translating, interpreting, describing and stating the main ideas. Translation, interpretation, extrapolation. Questions like: Compare the business and operational models of private clouds and hybrid clouds.
- *Application*: Using new knowledge. Solve problems in new situations by applying acquired knowledge, facts, techniques and rules in a different way. Questions like: What data analytics methods should be applied for specific data types analysis or for specific business processes and activities? Which big data services architecture is best suited for medium-sized research organisation or company, and why?
- *Analysis:* Examine and break information into parts by identifying motives or causes. Make inferences and find evidence to support generalisations. Analysis of elements, relationships, organisational principles. Questions like: What data analytics methods and services are required to support typical business processes of a web trading company? Give suggestions how these services can be implemented with the selected data analytics platform, including on-premises or outsourced to cloud. Provide references to support your statements.
- *Synthesis:* Compile information together in a different way by combining elements in a new pattern or proposing alternative solutions. Production of a unique communication, a plan or proposed set of operations, derivation of a set of abstract relations. Questions like: Describe the main steps and tasks for implementing data analytics and data management services for an example company or research organisation? What services and data analytics can be moved to clouds and which will remain at the enterprise premises and run by company's personnel?
- *Evaluation:* Present and defend opinions by making judgements about information, validity of ideas or quality of work based on a set of criteria. Judgements in terms of internal evidence or external criteria. Questions like: Do you think that implementing agile data-driven enterprise model creates benefits for enterprises, short term and long term?

Figure 4.1 illustrates Bloom's taxonomy learning levels. For instance, students start at the *knowledge* level when they can *name* and *identify* relevant technologies. They further move to *comprehension* level when they can *explain* how technologies work. They can then move to *application* level when they can *choose* right

Fig. 4.1 Simple Bloom's
taxonomy: learning levels
and action verbs

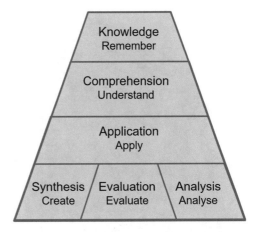

technology to *solve* a problem. Further they can progress to *analysis, synthesis* and
finally *evaluation* levels.

Figure 4.2 provides consolidated presentation of the Bloom's taxonomy [39, 40]
structure, attributes and action verbs that can be effectively used for designing
effective curricula and knowledge evaluation. When designing learning outcomes
for a course or programme, it is essential to ensure that all levels will be adequately
covered. Consideration of Bloom's taxonomy assists instructors both on the design
phase of a course or programme and during grading process. It is a reliable and
simple method to distinguish e.g. between familiarity with many concepts and being
able to use them in a practical setting.

Constructive Alignment and Problem-Based Learning The traditional and still
usual approach in science and engineering education is based on a behaviourist or
objectivist epistemology, in which the student is passively imparted with knowledge
by the teacher. Student's participation in the learning process is limited to
memorising schemes given by the instructor, which are assessed through instruments
such as examinations and quizzes that measure the degree of conformance to a norm
instead of actual competences [41]. In contrast, a constructivist epistemology puts
the student in the centre of the learning process as an active participant in
constructing knowledge [42].

Problem-Based Learning (PBL) [43, 44] is an alternative approach to instruction
based on providing student with a nontrivial problem to solve and guidance in
obtaining the necessary competencies. PBL is underlined by a constructivist epistemology that emphasises active student participation in the construction of their
knowledge from learning activities and motivating them through careful alignment
of evaluation activities, leading to a concept called constructive alignment described
by Biggs [45]. Ben-Ari [46] describes the applicability of constructivism to computer science education. Despite certain differences in epistemology between computer science and other sciences, constructivism is a useful approach to computer
science education.

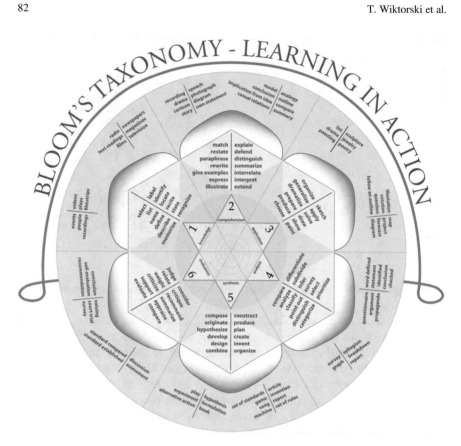

Fig. 4.2 Extended Bloom's taxonomy: consolidated presentation of learning levels, action verbs and associated learning instruments. https://en.wikipedia.org/wiki/Bloom%27s_taxonomy#/media/File:Blooms_rose.svg. It is CC BY SA 3.0

From the perspective of a whole education programme, constructive alignment and problem-based learning can be implemented in the form of project-based learning. In such a model, regular classes provide students with competences related to specific knowledge areas, while additional project classes allow us to establish a link between these competences. In addition, project classes provide an opportunity to reach higher levels of learning. An example of such approach on an institutional scale is University of Aalborg [47].

These education concepts provide guidance for further definition of learning outcomes and finally model curricula and can be used for the existing programmes evaluation.

Competence-Based Learning Model Competency-based learning (CBL) or competence-based education (CBE) also known as outcomes-based learning uses a different from the traditional education approach. Instead of focusing on how much time students spend learning a particular topic or concept (Carnegie unit credit hour, so called "sit time"), the CBL assesses whether students have mastered the given competencies, namely the knowledge, skills and abilities [34]. The learner (student or trainee) is evaluated on the specified (group of) competences, and only after mastering them they can move on to others. The CBL is also associated with more flexible study model for already working learners or those who undergo professional re-skilling or want to train for a new profession based on their existing experience, competences and skills. In this case, they can skip learning modules entirely if they can demonstrate required competences through the assessment system or formal testing.

The CBL can also allow the students to learn in their own pace, practising necessary skills as much as they need to achieve necessary mastery level. It works naturally with both individual self-study and with teacher or instructor supervised/facilitated study, so well suited for online and remote education, and for postgraduate education. CBL is also associated with such educational technologies and models as MOOCs, flipped classrooms, learning analytics and others targeting growing needs of lifelong learning and self-re-skilling dictated by current fast technologies development. The CBL programmes should offer the following features [48]:

- Self-pacing
- Modularisation
- Effective assessments
- Intentional and explicit learning objectives shared with the student
- Anytime/anywhere access to learning objects and resources
- Personalised, adaptive or differentiated instruction
- Learner supports through instructional advising or coaching

Although there are many examples of universities using CBL/CBE model, its practical implementation may create problems in some universities. Paper [27] by formulates the following principles that would allow integrating CBE into existing campus structures:

- The degree reflects robust and valid competencies.
- Students are able to learn at a variable pace and are supported in their learning.
- Effective learning resources are available any time and are reusable.
- Assessments are secure and reliable.

It is apparent that CBL is well suited for professional education and training of one of the EDISON target groups the self-made or practising data scientists. It is admitted [48] that the CBL was actually created to address needs of non-traditional students who cannot devote their full time to traditional academic study as well as effective model for companies to provide (re/up)-skilling their staff.

As stated above, the curriculum model has been built linking DS-BoK knowledge areas and MC-DS learning units for target competence groups. We explain now in more detail how it was done.

In general, a model curriculum can be regarded as a blueprint that can be used by educators and trainers to develop curricula at various educational institutions. There are several concepts that can guide the development of a curriculum like: alignment and coherence, scope, sequence, continuity and integration [9]. These 5 basic concepts help to develop a logically consistent curriculum, the components (courses and learning units) of which complement each other and are ordered in such a way that it forms a continuous, logical and progressive learning path. There are several common frameworks used to develop model curricula; some are subject or discipline centric while others are organised around concept and skills that are revised as we progress across the curriculum. In practice, model curricula should define either the time students must spend learning given topics (usually using credit units) or the outcome assessing whether students have mastered the given competences (knowledge, abilities and skills). The latter approach is known as competence-based education (CBE) or outcomes-based learning (OBL). In this case, well-defined learning outcomes for all academic activities or classes are specified, and students' progress is assessed against those learning outcomes.

The model curriculum is organised as core and elective topics, following the ACM definition [25]. Core topics are required for every data science programme while elective topics aim to cover in depth the knowledge on a specific area of data science. The last step identifies the learning outcomes associated with each core or elective topic.

The approach to defining the Data Science Model Curriculum follows a competence-based education model and can be summarised in the following five steps:

1. For each enumerated competence from CF-DS, define learning outcome according to knowledge or mastery level (defined as familiarity, usage, assessment for current MC-DS version).
2. Each knowledge area group of DS-BoK (that includes both KAGs from existing BoKs and those defined based on the ACM Classification Computer Science CCS2012) is mapped to existing academic subject classification groups that is primarily based on ACM CS2012 complemented with the domain- or technology-specific classifications such as BABOK, ACM-BOK, DAMA-BOK, PM-BOK and others to be defined by subject matter experts.
3. For each KAG or knowledge unit, specify related learning units defined according to academic subject classification or following current practices by universities.
4. For each learning unit, assign/suggest its category as core/mandatory (Tier 1 or Tier 2), elective or prerequisite.
5. For both core or elective, define a list of learning outcomes.

In the following, we compare mastery levels as used in the European Qualifications Framework (EQF) [8], The European e-Competence Framework, e-CFv3.0 [9], ACM/IEEE guidelines for computer science curriculum [25] and Bloom's taxonomy [39]. It leads to the definition of mastery levels (also called proficiency levels in e-CF) necessary to define learning outcomes in MC-DS. The e-CFv3.0 uses EQF for defining the proficiency level of knowledge and skills related to specific competences.

The European Qualification Framework (EQF) defines eight levels of knowledge achieved through stages of education. Level 6 is considered to be achieved through a bachelor's degree, level 7 through a master's degree and level 8 through a PhD degree. Levels 3-8 are mapped to 5 levels in e-CF dimension 3.

EQF descriptions provide reference both to actual levels of knowledge and to additional skills related to knowledge application, analysis, synthesis and evaluation. It is quite similar to Bloom's approach. At the same time, levels in EQF do not only correspond to higher levels of conceptualisation, but also to more specialised knowledge, experience and interpersonal skills related to people management and professional integrity and responsibility. e-CFv3.0 adds to its description of typical tasks regarding their complexity and autonomy. Therefore, higher levels of EQF and e-CFv3.0 should not just be seen directly as the same higher levels in Bloom. At the same time, higher levels in Bloom's taxonomy are necessary to move up in e-CFv3.0 and EQF.

EQF has 8 levels, e-CFv3.0 has 5 levels and Bloom's taxonomy has 6 levels. Designing LOs of whole programmes is a balance between precision and avoiding micromanagement of further definition of courses, especially when designing a guideline for programmes instead of a specific programme. It might be useful to limit the amount of levels on which LOs are considered. Such an approach is used in ACM-IEEE computer science and information technology curricula guidelines. Information technology guidelines [25] define the three levels as: emerging, developed and highly developed. Computer science guidelines [26] define the three levels as familiarity, usage and assessment. Bloom's taxonomy defines the six levels: knowledge, comprehension, application, analysis, synthesis and evaluation.

The three levels as used in ACM/IEEE computer science guidelines are of particular importance because significant parts of a related ACM/IEEE taxonomy and BoK are used in the definition of CF-DS and BoK-DS in EDSF. The verb usage is not fully consistent with the original Bloom's taxonomy or revised version, which is acknowledged in the document.

The comparison of the mastery levels definition used in EQF, e-CFv3.0, ACM-IEEE guidelines for computer science curriculum and Bloom's taxonomy is provided.

While not required in undergraduate curricula, the holistic definition covering all EQF, e-CF levels, would require also closer link between mastery levels and levels in Bloom's taxonomy (there may not be a direct mapping between mastery, proficiency or qualification levels and Bloom's taxonomy as it is primary defined for cognitive processes during learning in the time frame of curricula or individual courses, while proficiency level related to competences is the combination of knowledge and experience related to professional activity). At the same time, limitation to 3 levels should be maintained to preserve simplicity and compatibility. For the proposed MC-DS, we will use the following three levels: familiarity as understood by knowledge and comprehension in Bloom's taxonomy, usage as understood by application and analysis in Bloom's taxonomy, creation as understood by synthesis and evaluation in Bloom's taxonomy. We present the three levels again in this document for reference in Table 4.1. Action verbs were defined based

Table 4.1 Knowledge levels for learning outcomes in data science model curricula (MC-DS)

Level	Action verbs
Familiarity	Choose, classify, collect, compare, configure, contrast, define, demonstrate, describe, execute, explain, find, identify, illustrate, label, list, match, name, omit, operate, outline, recall, rephrase, show, summarise, tell, translate
Usage	Apply, analyse, build, construct, develop, examine, experiment with, identify, infer, inspect, model, motivate, organise, select, simplify, solve, survey, test for, visualise
Assessment	Adapt, assess, change, combine, compile, compose, conclude, criticise, create, decide, deduct, defend, design, discuss, determine, disprove, evaluate, imagine, improve, influence, invent, judge, justify, optimise, plan, predict, prioritise, prove, rate, recommend, solve

on the original and revised Bloom's taxonomy with adjustments tailored to data science curricula.

Now we define the *MC-DS learning units*. The MC-DS learning units (LU) or courses can be defined based on the knowledge area groups and knowledge units defined in the DS-BoK. The following Sect. 5 provides example defining courses or modules related to the DS-BoK knowledge area group on data science and analytics KAG1-DSDA and knowledge area group on data science engineering KAG2-DSENG. The individual learning units or courses are defined in accordance with the existing classification of academic disciplines, in particular the ACM CCS2012, and are verified with the existing offered courses at universities.

The proposed LUs are grouped according to ACM CCS2012 classification or DS-BoK knowledge groups/units that can be used as a context information for future data science curricula development, modification or enhancement with the linked courses and disciplines.

In the following sections, we present learning outcomes related to enumerated CF-DS competences and different knowledge/proficiency levels defined based on Bloom's taxonomy, with the general learning outcomes defined after CF-DS competences that are in most cases split into 3 knowledge levels and use specific verbs that reflect necessary comprehension or mastery level.

4.2.1 Data Analytics Learning Outcome (LO1-DA)

Learning outcome 1, Data analytics, DSDA. Its acronym is LO1-DA. The global Learning outcome (LO) of data analytics (DSDA-DA) is: use appropriate data analytics and statistical techniques on available data to discover new relations and deliver insights into research problem or organisational processes and support decision-making.

The learning outcomes (LOs) for the whole DSDA are denoted as **LO1-DA** and specified at three levels:

- Familiarity: Choose appropriate existing analytical method and operate existing tools to do specified data analysis. Present data in the required form.
- Usage: Develop data analysis application for specific datasets and tasks or processes. Identify necessary methods and use them in combination if necessary. Identify relations and provide consistent reports and visualisations.
- Assessment: Create formal model for the specific organisational tasks and processes and use it to discover hidden relations, propose optimisation and improvements. Develop new models and methods if necessary. Recommend and influence organisational improvement based on continuous data analysis.

The learning outcomes for specific DSDA competences are:

1. **LO1.01** based on DSDA01. Effectively use variety of data analytics techniques, such as machine learning (including supervised, unsupervised, semi-supervised learning), data mining, prescriptive and predictive analytics, for complex data analysis through the whole data life cycle.

   (a) Familiarity: Choose and execute existing data analytics and predictive analytics tools.
   (b) Usage: Identify existing requirements and develop predictive analysis tools.
   (c) Assessment: Design and evaluate predictive analysis tools to discover new relations.

2. **LO1.02** based on DSDA02. Apply designated quantitative techniques, including statistics, time series analysis, optimisation and simulation to deploy appropriate models for analysis and prediction.

   (a) Familiarity: Choose and execute standard methods from existing statistical libraries to provide overview.
   (b) Usage: Select most appropriate statistical techniques and model available data to deliver insights.
   (c) Assessment: Assess and optimise organisation processes using statistical techniques.

3. **LO1.03** based on DSDA03. Identify, extract and pull together available and pertinent heterogeneous data, including modern data sources such as social media data, open data, governmental data.

   (a) Familiarity: Operate tools for complex data handling.
   (b) Usage: Analyse available data sources and develop tools that work with complex datasets.
   (c) Assessment: Assess, adapt and combine data sources to improve analytics.

4. **LO1.04** based on DSDA04. Understand and use different performance and accuracy metrics for model validation in analytics projects, hypothesis testing and information retrieval.

   (a) Familiarity: Name and use basic performance assessment metrics and tools.

(b) Usage: Use multiple performance and accuracy metrics, select and use most appropriate for specific type of data analytics application.
(c) Assessment: Evaluate and recommend the most appropriate metrics, propose new for new applications.

5. **LO1.05** based on DSDA05. Develop required data analytics for organisational tasks, integrate data analytics and processing applications into organisation workflow and business processes to enable agile decision-making.

   (a) Familiarity: Define data elements necessary to develop specified data analytics.
   (b) Usage: Develop specialised analytics to enable decision-making.
   (c) Assessment: Design specialised analytics to improve decision-making.

6. **LO1.06** based on DSDA06. Visualise results of data analysis, design dashboard and use storytelling methods.

   (a) Familiarity: Choose and execute standard visualisation.
   (b) Usage: Build visualisations for complex and variable data.
   (c) Assessment: Create and optimise visualisations to influence executive decisions.

4.2.2 Data Engineering Learning Outcome (LO2-DSENG)

Learning outcome 2, Data engineering, DSENG. Its acronym is LO2-ENG. The global learning outcome (LO) of data engineering (DSENG) is: use engineering principles and modern computer technologies to research, design, implement new data analytics applications; develop experiments, processes, instruments, systems, infrastructures to support data handling during the whole data life cycle.

The learning outcomes (LOs) for the whole DSENG are denoted as **LO2-ENG** and specified at three levels:

- Familiarity: Identify and operate instruments and applications for data collection, analysis and management.
- Usage: Model problems and develop new instruments and applications for data collection, analysis and management following established engineering principles.
- Assessment: Evaluate instruments and applications to optimise data collection, analysis and management.

The learning outcomes for specific DSENG competences are:

1. **LO2.01** based on DSENG01. Use engineering principles (general and software) to research, design, develop and implement new instruments and applications for data collection, storage, analysis and visualisation.

(a) Familiarity: Choose potential technologies to develop, structure, instrument, machines, experiments, processes and systems.
(b) Usage: Model data analytics application to better develop suitable instruments, machines, experiments, processes and systems.
(c) Assessment: Create innovative solution to research and design data analytics.

2. **LO2.02** based on DSENG02. Develop and apply computational and data-driven solutions to domain-related problems using a wide range of data analytics platforms, with the special focus on big data technologies for large datasets and cloud-based data analytics platforms.

(a) Familiarity: Name computational solution and identify potential data analytics platform.
(b) Usage: Apply existing computational solutions to data analytic platform.
(c) Assessment: Adapt and optimise existing computational solutions to better fit to a given data analytics platform.

3. **LO2.03** based on DSENG03. Develop and prototype specialised data analysis applications, tools and supporting infrastructures for data-driven scientific, business or organisational workflow; use distributed, parallel, batch and streaming processing platforms, including online and cloud-based solutions for on-demand provisioned and scalable services.

(a) Familiarity: Identify a set of potential data analytics tools to fit specification.
(b) Usage: Survey various specialised data analytics tools and identify the best option.
(c) Assessment: Evaluate and recommend optimal data analytics tools to influence decision-making.

4. **LO2.04** based on DSENG04. Develop, deploy and operate large-scale data storage and processing solutions using different distributed and cloud-based platforms for storing data (e.g. data lakes, Hadoop, Hbase, Cassandra, MongoDB, Accumulo, DynamoDB, others).

(a) Familiarity: Find possible database solutions including both relational and non-relational databases.
(b) Usage: Model the problem to apply database technology.
(c) Assessment: Predict the difference in terms of performance between relational and non-relational databases and recommend a solution.

5. **LO2.05** based on DSENG05. Consistently apply data security mechanisms and controls at each stage of the data processing, including data anonymisation, privacy and IPR protection.

(a) Familiarity: Identify security issues related to reliable data access.
(b) Usage: Analyse security threats and solve them using known techniques.
(c) Assessment: Evaluate security threats and recommend adequate solutions.

6. **LO2.06** based on DSENG06. Design, build and operate relational and non-relational databases (SQL and NoSQL), integrate them with the modern Data Warehouse solutions, ensure effective ETL (Extract, Transform, Load), OLTP, OLAP processes for large datasets.

   (a) Familiarity: Define technical requirements for SQL/NoSQL databases, data warehouse technologies for data ingest.
   (b) Usage: Apply existing SQL/NoSQL databases, data warehouse technologies for creating data pipelines.
   (c) Assessment: Combine several techniques and optimise them to design new or custom environment to integrate existing DW and database technologies for new type of data and analytic applications.

4.2.3 Data Management Learning Outcome (LO3-DSDM)

Learning outcome 3, Data management, DSDM. Its acronym is LO3-DM. The global learning outcome (LO) of data science data management (DSDM) is: Develop and implement data management strategy for data collection, storage, preservation and availability for further processing.

The learning outcomes (Los) for the whole DSDM are denoted as **LO3-DM** and specified at three levels:

- Familiarity: Execute data strategy in the form of data management plan and illustrate how available software can help to promote data quality and accessibility.
- Usage: Develop components of data strategy and methods that improve quality, accessibility and publications of data.
- Assessment: Create data management plan aligned with the organisational needs, evaluate IPR and ethical issues.

The learning outcomes for specific DSDM competences are:

1. **LO3.01** based on DSDM01. Develop and implement data strategy, in particular, in the form of data management plan (DMP).

   (a) Familiarity: Explain and execute data strategy in the form of data management plan.
   (b) Usage: Develop components of data strategy in the form of data management plan.
   (c) Assessment: Assess various data strategies and create strategy, in the form of data management plan, aligned with organisational.

2. **LO3.02** based on DSDM02. Develop and implement relevant data models, including metadata.

   (a) Familiarity: Operate data models including metadata.

(b) Usage: Experiment with data models and model relevant metadata.

(c) Assessment: Evaluate and design data models, including metadata.

3. **LO3.03** based on DSDM03. Collect and integrate different data source and provide them for further analysis.

   (a) Familiarity: Collect different data sources.
   (b) Usage: Survey and visualise connection between different data sources.
   (c) Assessment: Compose different data sources to enable further analysis.

4. **LO3.04** based on DSDM04. Develop and maintain a historical data repository of analysis results (data provenance).

   (a) Familiarity: Operate a historical data repository.
   (b) Usage: Construct a historical data repository.
   (c) Assessment: Improve or design a historical data repository.

5. **LO3.05** based on DSDM05. Ensure data quality, accessibility, publications (data curation).

   (a) Familiarity: Illustrate how available software can help promote data quality, accessibility and publications.
   (b) Usage: Develop methods that improve quality, accessibility and publications of data.
   (c) Assessment: Improve quality, accessibility and publications.

6. **LO3.06** based on DSDM06. Manage IPR and ethical issues in data management.

   (a) Familiarity: Configure data management software to manage IPR and ethical issues.
   (b) Usage: Identify IPR and ethical issues in data repository.
   (c) Assessment: Evaluate IPR and ethical issues in data repository.

4.2.4   Research Methods and Project Management Learning Outcome, LO4-DSRMP

Learning outcome 4, Research methods and project management, DSRMP. Its acronym is LO4-RMP. The global learning outcome (LO) of DSRMP is: Create new understandings and capabilities by using the scientific method (hypothesis, test/ artefact, evaluation) or similar engineering methods to discover new approaches to create new knowledge and achieve research or organisational goals.

The learning outcomes (LOs) for the whole DSRMP are denoted as **LO4-RMP** and specified at three levels:

- Familiarity: Match elements of scientific or similar method and identify appropriate actions for organisational strategy to create new capabilities.

- Usage: Apply scientific or similar method and develop action plans to translate organisational strategies to create new capabilities.
- Assessment: Evaluate methodologies to optimise the development of organisational objectives.

The learning outcomes for specific DSRMP competences are:

1. **LO4.01** based on DSRMP01. Create new understandings by using the research methods (including hypothesis, artefact/experiment, evaluation) or similar engineering research and development methods.

   (a) Familiarity: Match elements of scientific or similar method to a given problem.
   (b) Usage: Apply scientific method to create new understandings and capabilities.
   (c) Assessment: Evaluate various methods and predict which method can optimise creation of new understandings and capabilities.

2. **LO4.02** based on DSRMP02. Direct systematic study towards understanding of the observable facts and discover new approaches to achieve research or organisational goals.

   (a) Familiarity: Choose observable facts from an existing study for a better understanding.
   (b) Usage: Apply systematic study towards a fuller knowledge or understanding of the observable facts.
   (c) Assessment: Combine several methods to discover new approaches to achieve organisational goals.

3. **LO4.03** based on DSRMP03. Analyse domain-related research process model, identify and analyse available data to identify research questions and/or organisational objectives and formulate sound hypothesis.

   (a) Familiarity: Formulate and test hypothesis for specified task or research question.
   (b) Usage: Create full experiment to test hypothesis for domain-specific task or experiment.
   (c) Assessment: Analyse domain-related models and propose analytics methods, suggest new data or improve quality of used data.

4. **LO4.03** based on DSRMP04. Undertake creative work, making systematic use of investigation or experimentation, to discover or revise knowledge of reality, and use this knowledge to devise new applications, contribute to the development of organisational objectives.

   (a) Familiarity: Show creativity under guidance of a senior staff in discovering and revising knowledge.
   (b) Usage: Develop creative solutions using systematic investigation or experimentation to revise and discover knowledge.

(c) Assessment: Adapt common systematic investigation to design and plan creative work to discover or revise knowledge.

5. **LO4.05** based on DSRMP05. Design experiments which include data collection (passive and active) for hypothesis testing and problem-solving.

(a) Familiarity: Illustrate outstanding ideas to solve complex problems.
(b) Usage: Identify non-standard solutions to solve complex problems.
(c) Assessment: Recommend cost-effective solution to a complex problem.

6. **LO4.06** based on DSRMP06. Develop and guide data-driven projects, including project planning, experiment design, data collection and handling.

(a) Familiarity: Identify appropriate actions for a given project plan or experiment.
(b) Usage: Develop actions and action plan to translate strategies into actionable plan.
(c) Assessment: Recommend effective action plans to translate strategies, suggest new data to improve effectiveness.

4.2.5 *Business Process Management Learning Outcome, LO5-BA*

Learning outcome 5, Business process management (DSBA). Its acronym is LO5-BA. The global learning outcome (LO) of business process management (DSBA) is: Use domain knowledge (scientific or business) to develop relevant data analytics applications and adopt general data science methods to domain-specific data types and presentations, data and process models, organisational roles and relations.

The learning outcomes (LOs) for the whole DSBA are denoted as **LO5-BA** and specified at three levels:

- Familiarity: Match elements of a mathematical framework to a given business problem and operate data support services for other organisational roles.
- Usage: Model business problems into an abstract mathematical framework and identify critical points which influence development of organisational objectives.
- Assessment: Evaluate various methods to predict which method can optimise solving business problems and recommend strategies that optimise the development of organisational objectives.

The learning outcomes for specific DSBA competences are:

1. **LO5.01** based on DSBA01. Analyse information needs, assess existing data and suggest/identify new data required for specific business context to achieve organisational goal, including using social network and open data sources.

- Familiarity: Match elements of a mathematical framework to a given business problem.
- Usage: Model an unstructured business problem into an abstract mathematical framework.
- Assessment: Evaluate various methods and predict which method can optimise solving business problems.

2. **LO5.02** based on DSBA02. Operationalise fuzzy concepts to enable key performance indicators measurement to validate the business analysis, identify and assess potential challenges.

   (a) Familiarity: Match data to specification of services.
   (b) Usage: Analyse services to develop data specification.
   (c) Assessment: Assess and improve use of data in services.

3. **LO5.03** based on DSBA03. Deliver business-focused analysis using appropriate BA/BI methods and tools, identify business impact from trends and make business case as a result of organisational data analysis and identified trends.

   (a) Familiarity: Identify appropriate actions for management and organisational decisions.
   (b) Usage: Identify critical points which influence development of organisational objectives.
   (c) Assessment: Recommend strategies that optimise the development of organisational objectives.

4. **LO5.04** based on DSBA04. Analyse opportunity and suggest the use of historical data available at organisation for organisational processes optimisation.

   (a) Familiarity: Operate data support services for other organisational roles.
   (b) Usage: Develop data support services for other organisational roles.
   (c) Assessment: Optimise data support services for other organisational.

5. **LO5.05** based on DSBA05. Analyse customer relations data to optimise/improve interacting with the specific user groups or in the specific business sectors.

   (a) Familiarity: Summarise customer data.
   (b) Usage: Survey and visualise customer data.
   (c) Assessment: Recommend actions based on data analysis to improve customer relations.

6. **LO5.06** based on DSBA06. Analyse multiple data sources for marketing purposes; identify effective marketing actions.

   (a) Familiarity: Access and use external open data and social network data.
   (b) Usage: Identify data that bring value to used analytics for marketing. Use cloud-based solutions.
   (c) Assessment: Suggest new marketing models based on existing and external data.

4.3   Data Science Model Curriculum

The proposed MC-DS intends to provide guidance to universities and training organisations in the construction of data science programmes and individual courses selection that are balanced according to the requirements elicited from the research and industry domains. MC-DS can be used for the assessment and improvement of existing data science programmes with respect to the knowledge areas and competence groups that are associated with specific professional profiles. When coupled with individual or group competence benchmarking, MC-DS can also be used for building individual training curricula and professional (self/up/re-) skilling for effective career management.

MC-DS follows the competence-based curriculum design approach grounded in the data science competences defined in CF-DS and correspondingly defined learning outcomes (LO). The DS-BoK provides the basis for structuring the proposed MC-DS by knowledge area groups (KAG) and knowledge areas (KA) defined in correspondence with the CF-DS competence groups and individual competences. MC-DS design supports design of programmes and courses that make use of best educational practices, such as constructive alignment, problem- and project-based learning and Bloom's taxonomy.

This section presents a short overview of the MC-DS organisation and its application to defining knowledge topics (knowledge units) and learning outcomes for two main knowledge area groups: data analytics and data engineering (DSDA and DSENG are the most distinguishing for the data science curricula, and in particular for the target professional profiles DSP04-DSP09). It also provides suggestions for ECTS points specification for main professional profiles group: data science professionals DSP04–DSP09, which will be described in the next Chap. 5 "Data Science Professional Profiles".

4.3.1   Organisation and Application of Model Curriculum

MC-DS organisation is based on Data Science Competence Framework, Body of Knowledge and Professional Profiles. For each enumerated competence, MC-DS defines learning outcome according to knowledge or mastery level (defined as familiarity, usage, assessment). Each knowledge area group of DS-BoK is mapped to existing academic subject classification groups that is primarily based on ACM Classification Computer Science CCS2012 complemented with the domain- or technology-specific classifications such as defined in the existing BoK's ACM CS-BOK, BABOK, SWEBOK, DM-BoK, PM-BOK and others that should to be defined by subject matter experts. For each KAG, MC-DS specifies learning outcomes and mastery levels following Bloom's taxonomy verb usage. Learning outcomes are also linked to a set of learning units, which are examples of practical application of knowledge units. ECTS points are provided for professional profile

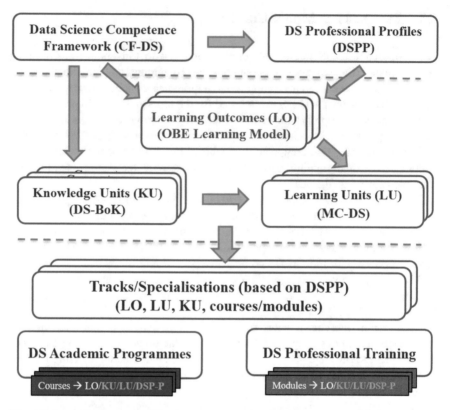

Fig. 4.3 Interaction between different components of EDSF when using model curriculum for defining academic of professional training programme for target professional group

groups and divided into Tier 1, Tier 2, elective and prerequisite categories to help create detailed tracks and specialisations for academic programmes and professional training.

Figure 4.3 illustrates the relation between different EDSF components when defining specific academic or professional training programme that can be tailored for specific target data science professional group.

Figure 4.4 illustrates a general approach to application of the model curriculum to create an educational programme. The work starts by deciding on a target Data Science Professional Profiles group the programme should cover and the level of the programme, usually bachelor's or master's. These elements allow to identify a set of competencies to be address in the programme. To identify relevant knowledge nits and to what extent they should be covered in the new programme, the programme designer can consult tables with ECTS point, which are defined for each professional profile. ECTS points specifications include a degree of flexibility to adjust to the particular needs. For each knowledge area, MC-DS defines a set of knowledge units based on BoK and a set of learning outcomes based on competence framework. Topics and learning outcomes become a base for definition of new courses or use of

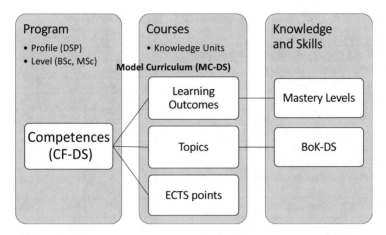

Fig. 4.4 Visualisation of model curriculum application for programmes and courses

existing courses. It is important to note that when designing a specific course, it may include elements from several knowledge areas to ensure consistency of the whole data science programme.

Adjustment of learning outcomes levels for different proficiency levels can be done based on the full list of learning outcomes for all CF-DS competences and for all mastery/proficiency levels provided in Sects. 4.2.1–4.2.5 of this document. Learning outcomes can repeat between subgroups within the same KAG, but adjusted to a specific course and topics context.

4.3.2 Assignments of European Credit Transfer and Accumulation System points to Competence Groups and Knowledge Areas

This section presents an example, European Credit Transfer and Accumulation System (ECTS) points specification for main professional profile group: data science professionals (DSP04-DSP09). Table 4.2 contains example specification for a programme on a Bachelor's level, while Table 4.3 provides example specification for a programme on master's level.

Points for each knowledge area are divided into four categories: Tier 1, Tier 2, elective and prerequisite. For each programme 100% of Tier 1 should be covered, 80% of Tier 2 and 50% of elective, with minor adjustments if necessary.

Such system ensures that each programme based on MC-DS covers basic competence and knowledge, but at the same time allowing for a necessary degree of flexibility. No prerequisites are expected for a bachelor's programme, while for a master's programme we set prerequisite at around 50% of combined Tier 1 and Tier 2. The goal is to ensure that students entering a programme have at least basic

Table 4.2 ECTS credit points for BSc programme for profiles DSP04-09

Course related to DS-BoK KA	Tier1	Tier2	Elective	Prerequisite
DSDA/SMA (Statistical methods and data analysis)	7	4	6	NA
DSDA/ML (Machine learning)	9	8	8	NA
DSDA/DM (Data mining)	5	4	3	NA
DSDA/TDM (Text data mining)	4	3	3	NA
DSDA/PA (Predictive analytics)	6	7	6	NA
DSDA/MSO (Modelling, simulation and optimisation)	5	3	4	NA
DSENG/BDI (Big data infrastructure and technologies)	4	3	4	NA
DSENG/IPDS (Infrastructure and platforms for data science)	8	5	4	NA
DSENG/CCT (Cloud computing technologies for BD and DA)	6	5	5	NA
DSENG/SEC (Data and applications security)	2	2	2	NA
DSENG/BDSE (Big data systems organisation and engineering)	9	5	5	NA
DSENG/DSAD (Data science/big data application design)	9	5	5	NA
DSENG/SE (Information systems)	4	6	5	NA

Table 4.3 ECTS credit points for MSc programme for profiles DSP04-09

Course related to DS-BoK KA	Tier1	Tier2	Elective	Prerequisite
DSDA/SMA (Statistical methods and data analysis)	6	2	4	6
DSDA/ML (Machine learning)	6	5	5	9
DSDA/DM (Data mining)	4	2	4	5
DSDA/TDM (Text data mining)	3	2	4	4
DSDA/PA (Predictive analytics)	4	4	4	7
DSDA/MSO (Modelling, simulation and optimisation)	2	2	4	4
DSENG/BDI (Big data infrastructure and technologies)	3	3	3	4
DSENG/IPDS (Infrastructure and platforms for data science)	5	3	4	7
DSENG/CCT (Cloud computing technologies for BD and DA)	5	3	4	6
DSENG/SEC (Data and applications security)	1	2	2	2
DSENG/BDSE (Big data systems organisation and engineering)	5	3	4	7
DSENG/DSAD (Data science/big data application design)	5	3	4	7
DSENG/SE (Information systems)	2	3	3	5

competences necessary to succeed in master's education, but at the same time it allows students from relatively wide set of backgrounds to participate.

Students who do not possess the required competences should be able to make up the difference by engaging in additional courses or bootcamps. In case, programme wants to accept student with a different profile, e.g. pure computer science or pure statistics, we recommend that distribution of points in the programme is adjusted to balance that. For instance, students with a BSc in computer science come with a strong background in software development and databases, but limited knowledge of statistics. In such a case, ECTS points should be moved between these areas.

Suggested ECTS specification for data analytics and data science engineering knowledge area groups is presented here. Points for data management and research methods will be presented in the MC-DS development. They should complement the ECTS points from two groups presented here to provide 180 ECTS for bachelor's programmes and 120 ECTS for master's programmes. Actual curriculum design and courses selection should be done by the programme coordinator at the hosting organisations to reach total required number of credits and contain course from Tier 1, Tier 2 and elective.

4.3.3   Data Analytics-Related Courses

Data analytics knowledge group builds the ability to use appropriate statistical and data analytics techniques on available data to deliver insights and discover information, provide recommendations and support decision-making. It includes knowledge areas that cover data mining, supervised and unsupervised machine learning, statistical modelling and predictive analytics.

The following are commonly defined data analytics knowledge areas group (KAG01-DSDA):

- KA01.01 (DSDA.01/SMA) Statistical methods, including descriptive statistics, exploratory data analysis (EDA) focused on discovering new features in the data and confirmatory data analysis (CDA) dealing with validating formulated hypotheses
- KA01.02 (DSDA.02/ML) Machine learning and related methods for information search, image recognition, decision support, classification
- KA01.03 (DSDA.03/DM) Data mining as a particular data analysis technique that focuses on modelling and knowledge discovery for predictive rather than purely descriptive purposes
- KA01.04 (DSDA.04/TDM) Text analytics applies statistical, linguistic and structural techniques to extract and classify information from textual sources, a species of unstructured data
- KA01.05 (DSDA.05/PA) Predictive analytics focuses on application of statistical models for predictive forecasting or classification

- KA01.06 (DSDA.06/MSO) Computational modelling, simulation and optimisation

The proposed topics for the courses and learning outcomes for each one of them are specified in the following subsections.

4.3.3.1 DSDA.01/SMA. Statistical Methods and Data Analysis

Statistics and probability theory are foundational components of data analytics and constitute a significant part of a data science competences and knowledge. This module provides insights into major statistical and data analytics paradigms and schools of thought. They can be taught separately or as a part of other data analytics-related modules or courses.

- Topics:

  - Statistical paradigms (regression, time series, dimensionality, clusters)
  - Probabilistic representations (causal networks, Bayesian analysis, Markov nets)
  - Frequentist and Bayesian statistics
  - Exploratory and confirmatory data analysis
  - Information theory
  - Graph theory

- Learning outcomes

  - Choose and execute standard methods from existing statistical libraries to provide overview (LODA.02 L1)
  - Select most appropriate statistical techniques and model available data to deliver insights (LODA.02 L2)
  - Identify requirements and develop analysis approaches (LODA.01 L2)
  - Assess and optimise organisation processes using statistical techniques and simulation (LODA.02 L3)

4.3.3.2 DSDA.02/ML. Machine Learning

Data Scientists have a wide range of ready machine learning libraries available. Nevertheless, they also need to go beyond simple application of algorithms to achieve expected results. New problems they face might require an in-depth understanding of theoretical underpinning of both simple and advanced algorithms. This module covers the use, analyse and design of machine learning algorithms.

- Topics:

- Machine learning theory (supervised, unsupervised, reinforced learning, deep learning, kernel methods, Markov decision processes)
- Design and analysis of algorithms (graph algorithms, data structures design and analysis, online algorithms, Bloom filters and hashing, MapReduce algorithms)
- Game theory and mechanism design
- Classification methods
- Ensemble methods
- Cross-validation

- Learning outcomes

  - Choose and execute existing analytic techniques and tools (LODA.01 L1)
  - Identify requirements and develop analysis approaches (LODA.01 L2)
  - Develop specialised analytics to enable agile decision-making and integrate them into organisational workflows (L0DA.05 L2)
  - Design and evaluate analysis techniques and tools to discover new relations (LODA.01 L3)

4.3.3.3   DSDA.03/DM. Data Mining

Mathematical and theoretical aspects of data analytics must be implemented in a computational form appropriate for both problem at hand and data size. This module builds familiarity with most relevant data mining algorithms and related methods for knowledge representation and reasoning.

- Topics:

  - Data mining and knowledge discovery
  - Knowledge representation and reasoning
  - CRISP-DM and data mining stages
  - Anomaly detection
  - Time series analysis
  - Feature selection, a priori algorithm
  - Graph data analytics

- Learning outcomes

  - Choose and execute standard methods from statistical libraries to provide overview (LODA.02 L1)
  - Select most appropriate statistical techniques and model available data to deliver insights (LODA.02 L2)
  - Analyse available data sources and develop tool that works with complex datasets (LODA.03 L2)

- Develop specialised analytics to enable agile decision-making and integrate them into organisational workflows (LODA.05 L2)
- Evaluate and recommend data analytics organisational strategy (LODA.05 L3)

4.3.3.4   DSDA.04/TDM. Text Data Mining

Text data mining can be considered a subset of data mining, but it is worth a separate consideration due to the amount of text data available and particular methods developed over the years to analyse it.

- Topics

  - Text analytics including statistical, linguistic and structural techniques to analyse structured and unstructured data
  - Data mining and text analytics
  - Natural language processing
  - Predictive models for text
  - Retrieval and clustering of documents
  - Information extraction
  - Sentiments analysis

- Learning outcomes

  - Choose and execute standard methods from statistical libraries to provide overview (LODA.02 L1)
  - Analyse available data sources and develop tool that works with complex datasets (LODA.03 L2)
  - Evaluate and recommend data analytics organisational strategy (LODA.05 L3)

4.3.3.5   DSDA.05/PA. Predictive Analytics

Predictive analytics are a commonly used to foresee future events in order to avoid them or act ahead. This module covers both traditional approaches based on time series and newer approaches based on deep learning. Anomaly detection is a particular focus since it is one of most common application areas.

- Topics

  - Predictive modelling and analytics
  - Inferential and predictive statistics
  - Machine learning for predictive analytics
  - Regression and multi analysis
  - Generalised linear models

- Time series analysis and forecasting
- Deploying and refining predictive models

• Learning outcomes

  - Choose and execute existing analytic techniques and tools (LODA.01 L1)
  - Identify requirements and develop analysis approaches (LODA.01 L2)
  - Create stories and optimise visualisations to influence executive decisions (LODA.06 L3)

4.3.3.6   DSDA.06/MSO. Computational Modelling, Simulation and Optimisation

Modelling and simulation are essential approaches to handle the complexity of some systems and event chains. This module provides an introduction in both theoretical and practical aspects of model development and simulation techniques.

• Topics:

  - Modelling and simulation theory and techniques (general and domain oriented)
  - Operations research and optimisation
  - Large-scale modelling and simulation systems
  - Network optimisation
  - Risk simulation and queuing

• Learning outcomes

  - Describe and execute different performance and accuracy metrics (LODA.04 L1)
  - Compare and choose performance and accuracy metrics (LODA.04 L2)
  - Assess and optimise organisation processes using statistical techniques and simulation (LODA.02 L3)

4.3.4   Data Engineering-Related Courses

Data science engineering knowledge group builds the ability to use engineering principles to research, design, develop and implement new instruments and applications for data collection, analysis and management. It includes knowledge areas that cover software and infrastructure engineering, manipulating and analysing complex, high-volume, high-dimensionality data, structured and unstructured data, cloud-based data storage and data management.

Data science engineering includes software development, infrastructure operations, and algorithms design with the goal to support big data and data science applications in and outside the cloud. The following are commonly defined data science engineering knowledge areas group (KAG02-DSENG):

- KA02.01 (DSENG.01/BDIT) Big data infrastructure and technologies, including NOSQL databases, platforms for big data deployment and technologies for large-scale storage
- KA02.02 (DSENG.02/DSIAPP) Infrastructure and platforms for data science applications, including typical frameworks such as Spark and Hadoop, data processing models and consideration of common data inputs at scale
- KA02.03 (DSENG.03/CCT) Cloud computing technologies for big data and data analytics
- KA02.04 (DSENG.04/SEC) Data and applications security, accountability, certification and compliance
- KA02.05 (DSENG.05/BDSE) Big data systems organisation and engineering, including approached to big data analysis and common MapReduce algorithms
- KA02.06 (DSENG.06/DSAPPD) Data science (big data) application design, including languages for big data (Python, R), tools and models for data presentation and visualisation
- KA02.07 (DSENG.07/IS) Information systems, to support data-driven decision-making, with a focus on data warehouse and data centres.

The proposed topics for the courses and learning outcomes for each one of them are specified in the following subsections.

4.3.4.1 DSENG.01/BDIT. Big Data Infrastructure and Technologies

Big data infrastructures and technologies drive many of the data science applications. Systems and platforms behind big data differ significantly from traditional ones due to specific challenges of volume, velocity and variety of data. This module addresses these aspects with a focus on underlying storage technologies and distributed architectures.

- Topics:
    - Big data cloud platforms (Azure, AWS)
    - Approaches to data ingestion at scale
    - Parallel and distributed computer architectures (cloud computing, client/server, grid)
    - Large-scale storage systems, SQL and NoSQL databases
    - Computer networks architectures and protocols
    - Storage for big data infrastructures and high-performance computing (HDFS, Ceph)
- Learning outcomes

- Find possible data storage and processing solutions including both traditional and NOSQL databases (LOENG.06 L1)
- Survey various specialised data-driven tools and identify the best option (LOENG.03 L2)
- Evaluate the difference in performance between various distribute and cloud-based platforms and recommend a solution (LOENG.01 L3)

4.3.4.2   DSENG.02/DSIAPP. Infrastructure and Platforms for Data Science Applications

Deployment of data science applications is usually tied to one of most common platforms, such as Hadoop or Spark, hosted either on private or public cloud. The application must be also tied to a whole data processing pipeline including ingestion and storage. This module covers these aspects with an additional focus on handling most common types of data inputs at scale.

- Topics:

  - Big data frameworks (Hadoop, Spark, Cloudera, others)
  - Big data infrastructures (ingestion, storage, streaming, enabling analytics, Lambda Architecture)
  - Data processing models (batch, streaming, parallelism)
  - Large-scale data storage and management (data inputs: graph, text, image, table, time series)

- Learning outcomes

  - Define technical requirements for new distributed and cloud-based application for a given high-level design (LOENG.04 L1)
  - Apply existing data-driven solutions to data analytic platform (LOENG.02 L2)
  - Evaluate the difference in performance between various distribute and cloud-based platforms and recommend a solution (LOENG.04 L3)

4.3.4.3   DSENG.03/CCT. Cloud Computing Technologies for Big Data and Data Analytics

Cloud computing technologies are the most common way to deploy big data and data analytics applications. This module provides an introduction to various levels of cloud computing services, such as IaaS or PaaS on practical examples. It is also important to consider both private and public cloud.

- Topics

- Cloud computing architecture and services
- Cloud computing engineering (design, management, operation)
- Cloud-enabled applications development (IaaS, PaaS, SaaS, autoscaling)
- Capex vs Opex consideration

- Learning outcomes

  - Choose potential technologies to implement new applications for data collection and storage (LOENG.01 L1)
  - Model a problem to apply distributed and cloud-based platforms (LOENG.04 L2)
  - Evaluate the difference in performance between various distribute and cloud-based platforms and recommend a solution (LOENG.04 L3)

4.3.4.4   DSENG.04/SEC. Data and Applications Security

Data scientists should have a general understanding of data and application security aspects in order to properly plan and execute data-driven processing in the organisation. This module provides an overview of the most important aspects, including sometime omitted concepts of accountability, compliance and certification.

- Topics

  - Data security, accountability, protection
  - Blockchain and corresponding infrastructure
  - Access control and identity management
  - Compliance and certification
  - Data anonymisation and privacy

- Learning outcomes

  - Identify security issues related to reliable data access (LOENG.05 L1)
  - Analyse security threats and solve them using known techniques (LOENG.05 L2)

4.3.4.5   DSENG.05/BDSE. Big Data Systems Organisation and Engineering

Systems and platforms behind big data differ significantly from traditional ones due to specific challenges of volume, velocity and variety of data. They require specialised approaches to data processing and algorithm engineering. This module addresses aspects both in general and based on common MapReduce algorithms.

- Topics

  - Big data frameworks (Hadoop, Spark, Cloudera, others)
  - Algorithms for large-scale data processing
  - Methods for preprocessing data implemented in MapReduce, including problems of correct data splitting in clusters
  - Approaches to big data analysis (functional abstraction for data processing, MapReduce, Lambda Architecture)
  - Algorithms for visualisation of large datasets, including subsampling with different distributions
  - Big data systems for applications domains

- Learning outcomes

  - Choose potential technologies to implement new applications for data collection and storage (LOENG.01 L1)
  - Find possible data storage and processing solutions including both traditional and NOSQL databases (LOENG.06 L1)
  - Model data-driven application following engineering principles (LOENG.01 L2)
  - Adapt and optimise existing data-driven solutions to better fit to a given data analytics platform (LOENG.02 L3)

4.3.4.6   DSENG.06/DSAPPD. Data Science (Big Data) Application Design

Data Scientists are often tasked with developing new applications and systems. Certain languages and tools are more suitable in a data scientific context than others. This module covers most common languages for data science and big data processing together with most common tools for data presentation.

- Topics:

  - Languages for big data (Python, R)
  - Tools and models for data presentation and visualisation (Jupyter, Zeppelin)
  - Software requirements and design
  - Software engineering models and methods
  - Software quality assurance
  - Agile development methods, platforms, tools
  - DevOps and continuous deployment and improvement paradigm

- Learning outcomes

  - Identify a set of potential data analytics tools to fit specification (LOENG.03 L1)

- Define technical requirements for new distributed and cloud-based application for a given high-level design (LOENG.06 L1)
- Model data-driven application following engineering principles (LOENG.01 L2)
- Apply existing techniques to develop new data analytics applications (LOENG.02 L2)
- Combine several techniques and optimise them to design new data analytic applications (LOENG.06 L3)

4.3.4.7   DSENG.07/IS. Information Systems

All organisations rely on some form of information systems to preserve knowledge and drive decision processes. This module focuses on basics of well-established data warehouse, expert systems and decision support systems. Big data influence on such systems is also of interest, but related technical details are covered by other KAs.

- Topics:

  - Decision support systems
  - Data warehousing and expert systems
  - Enterprise information systems (data centres, intra/extra-net)
  - Multimedia information systems

- Learning outcomes

  - Identify a set of potential data-driven tools to fit specification (LOENG.03 L1)
  - Model the problem to apply traditional or NOSQL database technology (LOENG.06 L2)
  - Evaluate and recommend optimal data-driven tools to influence decision-making (LOENG.03 L3)

Chapter 5
Data Science Professional Profiles

Yuri Demchenko and Juan J. Cuadrado-Gallego

This chapter treats all the aspects related to defining the Data Science Professional Profiles that can be also called data-related occupations family. The proposed occupations are placed in four top classification groups: *managers*, for managerial roles; *professionals*, for applications developers and for infrastructure engineers; *technicians and associate professionals*, for operators and technicians; and *clerical support workers*, for data curators and stewards.

The fifth chapter, "Data Science Professional Profiles", has three sections with the following structure: Sect. 5.1 presents an overview of existing frameworks for Information and Communication Technologies and Data Science Professional Profiles definition; Sect. 5.2 presents the definition of the Data Science Professional Profiles; and finally Sect. 5.3 presents a comparison of competence, proficiency and qualifications levels for professional profiles in data science.

5.1 Overview of Existing Frameworks for Information and Communication Technologies and Data Science Professional Profiles Definition

This section provides a brief overview of existing standard and commonly accepted frameworks for defining professional profiles for general information communications technologies (ICT) occupations and currently defined data handling-related professions.

Y. Demchenko
Universiteit van Amsterdam, Amsterdam, The Netherlands
e-mail: y.demchenko@uva.nl

J. J. Cuadrado-Gallego (✉)
Department of Computer Science, University of Alcalá, Madrid, Spain
e-mail: jjcg@uah.es

© Springer Nature Switzerland AG 2020
J. J. Cuadrado-Gallego, Y. Demchenko (eds.), *The Data Science Framework*,
https://doi.org/10.1007/978-3-030-51023-7_5

5.1.1   *Information and Communications Technologies Professional Profiles (EC CWA 16458)*

The European Commission ICT Professional Profiles, European Committee for Standardization, CEN Workshop Agreement, CWA 16458, was created to provide the basis for compatible ICT profiles definition by organisations and the basis for defining new profiles by European stakeholders [11].

The CWA defines 23 main ICT profiles the most widely used by organisations by defining organisational roles for ICT workers, which are grouped into the six ICT profile families:

1. Business management
2. Technical management
3. Design
4. Development
5. Service and operation
6. Support

The European ICT profile descriptions are reduced to core components and constructed to clearly differentiate profiles from each other. Further context-specific elements can be added to the profiles according to the specific environments in which the profiles are to be integrated. Figure 5.1 illustrates six ICT profile families and related main profiles which are non-exhaustive.

The 23 profiles constructed in CWA combined with e-competences from the e-CF3.0 [9] provide a pool for the development of tailored profiles that may be developed by European ICT sector players in specific contexts and with higher levels of granularity. The 23 profiles cover the full ICT business process; positioning them into the e-CF dimension 1 demonstrates this.

Figure 5.2 illustrates mapping between CWA families and e-CF3.0 competence areas and also CWA ICT profiles allocation to families and competence areas

5.1.2   *European Skills, Competences, Occupations and Qualifications Framework and Platform (EC ESCO)*

The Commission services launched the ESCO [7] project in 2010 with an open stakeholder consultation. Currently, DG Employment, Social Affairs and Inclusion coordinates the development of ESCO with the support by the European Centre for the Development of Vocational Training (Cedefop). Stakeholders are closely involved in the development and dissemination of ESCO.

The ESCO classification identifies and categorises skills, competences, qualifications and occupations relevant for the EU labour market and education and training. It systematically shows the relationships between the different concepts. ESCO has been developed in an open IT format, is available for use free of charge by everyone and can be accessed via the ESCO portal.

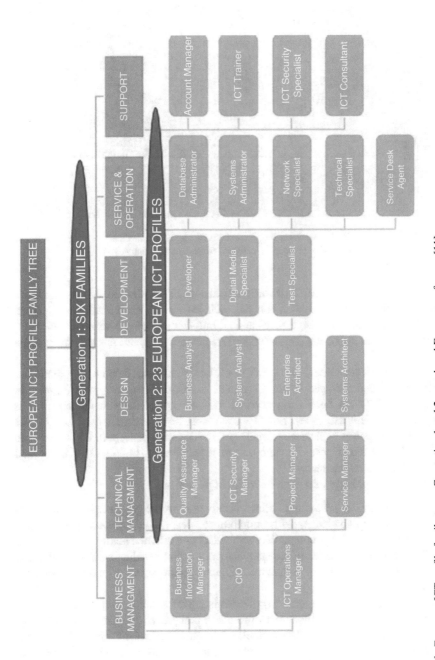

Fig. 5.1 European ICT profile family tree—Generation 1 and 2 as a shared European reference [11]

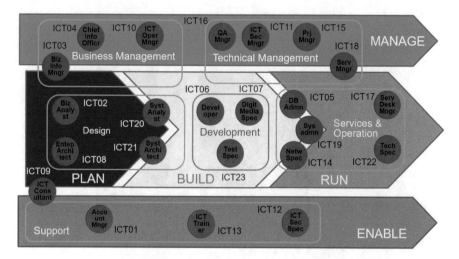

Fig. 5.2 European ICT Professional Profiles structured by six families and positioned within the ICT business process (e-CF dimension 1) (adopted from [9] and extended)

The first version of ESCO v0 was published on 23 October 2013. This version is based on the EURES classification but includes an enhanced semantic structure, cross-sector skills and competences and an initial small sample of qualifications. It includes the results of the Cross-Sector Reference Group, but not yet any sectoral updates.

The first full version of ESCO, ESCO v1, has been released on 28 July 2017. The release has been followed by the Conference *ESCO: Connecting people and jobs* whereby ESCO has been showcased in concrete applications to demonstrate its value in different use cases. ESCO v1 contains 2942 occupations, 13,485 skills and competences as well as some qualifications. As each concept in ESCO exists in all 26 ESCO languages, this amounts to more than +514,000 skills terms and +350,000 occupations terms [7].

ESCO is organised into three interrelated pillars:

- The *occupations pillar* that defines occupation profile which refers to essential and optional knowledge, skills and competences; higher level occupations classification in ESCO is compatible with ISCO as explained below.
- The *knowledge*, *skills* and *competences* pillar that besides the terms definition defines their reusability as transversal, cross-sector, sector-specific, occupation-specific.
- The *qualifications* pillar aims to collect existing information on qualifications. The final objective of the pillar is to provide a comprehensive list of qualifications relevant for the European labour market. The qualifications pillar in ESCO contains a small sample list of qualifications regulated at European level, international qualifications and certificates and licences linked to tasks, technologies, occupations or sectors. The qualifications pillar of ESCO is developed in full compliance and complementarity with the European Qualifications Framework (EQF) [9].

The following definitions are used in the document:

- *Knowledge*: The body of facts, principles, theories and practices that is related to a field of work or study. Knowledge is described as theoretical and/or factual and is the outcome of the assimilation of information through learning.
- *Skill*: The ability to apply knowledge and use know-how to complete tasks and solve problems. Skills are described as cognitive (involving the use of logical, intuitive and creative thinking) or practical (involving manual dexterity and the use of methods, materials, tools and instruments).
- *Competence*: The proven ability to use knowledge, skills and personal, social and/or methodological abilities, in work or study situations, and in professional and personal development.

Moreover, transversal skills and competences are organised into a hierarchical structure with the following five headings:

- Thinking
- Language
- Application of knowledge
- Social interaction
- Attitudes and values

ESCO Strategic Framework defines relation of ESCO with other European initiatives and standards; in particular ESCO will be used within the EURES network of employment services in order to exchange job vacancies and CVs between member states and with the Commission (Table 5.1).

In the following, the related occupations with data science extracted from the ESCO classification together with related hierarchies are presented. That is included for reference purposes to present the ESCO top level occupations classification. The order from top to down is top hierarchy (TH), hierarchy (H), skills/competence group (S/C) and occupations (O).

- **Managers (TH)**

    - Production and specialised services managers (H)
        Information and communications technology service managers (S/C)
        Access supervisor, data processing/IT (O)

- **Professionals (TH)**

    - Information and communications technology professionals (H)
    - Database and network professionals (H)
    - Database and network professionals not elsewhere classified (S/C)

        Security director (data processing/IT) (O)
        Security analyst (data processing/IT) (O)
        Supervisor (data processing) (O)
        Data processing investigator (O)

Table 5.1 ESCO top occupation hierarchy

ESCO occupations top level hierarchy
Armed forces occupations
Clerical support workers[a]
Numerical and material recording clerks
Other clerical support workers
Customer services clerks
General and keyboard clerks
Craft and related trade workers
Elementary occupations
Managers
Administrative and commercial managers
Chief executives, senior officials and legislators
Hospitality, retail and other services managers
Production and specialised services managers
Plant and machine operators and assemblers
Professionals
Teaching professionals
Science and engineering professionals
Health professionals
Legal, social and cultural professionals
Business and administration professionals
Information and communications technology professionals
Service and sales workers
Skilled agricultural, forestry and fishery workers
Technicians and associate professionals
Health associate professionals
Information and communications technicians
Legal, social, cultural and related associate professionals
Science and engineering associate professionals
Business and administration associate professionals

[a]The highlighted bold font indicates which ESCO taxonomy groups are identified for proposed extension with the data science occupations

- Database designers and administrators (S/C)

  Data recorder (O)
  Operations manager (data processing) (O)
  Data processing manager (O)
  Data processing analyst (O)
  Data processing supervisor (O)

- Systems administrators (S/C)

  Data processing consultant (O)

- Software and applications developers and analysts (H)

- Systems analysts (S/C)

  Data processing strategist (O)

- **Technicians and associate professionals (TH)**

  - Information and communications technicians (H)
  - Information and communications technology operations and user support technicians (H)
  - Information and communications technology operations technicians (S/C)

    Operations technician (data processing) (O)

5.1.3   International Standard Classification of Occupations (ISCO)

The International Standard Classification of Occupations (ISCO) [49, 50] is a tool for organising jobs into a clearly defined set of groups according to the tasks and duties undertaken in the job; its main aims are to provide:

- A basis for the international reporting, comparison and exchange of statistical and administrative data about occupations
- A model for the development of national and regional classifications of occupations
- A system that can be used directly in countries that have not developed their own national classifications

It is intended for use in statistical applications and in a variety of client-oriented applications. Client-oriented applications include the matching of job seekers with job vacancies, the management of short- or long-term migration of workers between countries and the development of vocational training programmes and guidance.

ISCO is a four-level classification of occupation groups managed by the International Labour Organization (ILO). Its structure follows a grouping by education level. The two latest versions of ISCO are ISCO-88 (dating from 1988) and ISCO-08 (dating from 2008). Many current national occupational classifications are based on one of these ISCO versions.

ISCO 08 groups jobs together in occupations and more aggregate groups mainly on the basis of the similarity of skills required to fulfil the tasks and duties of the jobs. Two dimensions of the skill concept are used in the definition of ISCO 88 groups:

- Skill level, which is a function of the range and complexity of the tasks involved, where the complexity of tasks has priority over the range.
- Skill specialisation, which reflects type of knowledge applied, tools and equipment used, materials worked on, or with, and the nature of the goods and services produced. It should be emphasised that the focus in ISCO 88 is on the skills required to carry out the tasks and duties of an occupation and not on whether a worker in a particular occupation is more or less skilled than another worker in the same or other occupations.

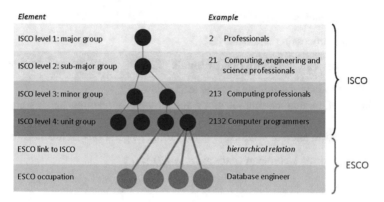

Fig. 5.3 Relation between ESCO and ISCO: ISCO provides 4 top hierarchy levels in ESCO occupations pillar [49]

In ESCO [7], each occupation is mapped to exactly one ISCO-08 code. ISCO-08 can therefore be used as a hierarchical structure for the occupations pillar in ESCO. ISCO-08 provides the top four levels for the occupations pillar. ESCO occupations are located at level 5 and lower. Figure 5.3 illustrates the role of ISCO 08 in the hierarchical structure of the ESCO occupations pillar.

Since ISCO is a statistical classification, its occupation groups do not overlap. Each ESCO occupation is therefore mapped to only one ISCO unit group. It follows from this structure that ESCO occupation concepts can be equal to or narrower than ISCO unit groups, but not broader. The result is a strictly mono-hierarchical structure where each element at level 2 or lower has exactly one parent. A few groups of ISCO-08 do not contain ESCO occupations. These are usually occupation groups with no economic activity in the EU, such as "water and firewood collectors".

5.2 Data Science Professional Profiles

This section presents initial results on defining the Data Science Professional Profiles that can be also called data-related occupations family. They are defined as an extension to the ESCO occupations taxonomy. The proposed new occupations are placed in four top classification groups:

- *Managers*, for managerial roles
- *Professionals*, for applications developers and for infrastructure engineers
- *Technicians and associate professionals*, for operators and technicians
- *Clerical support workers*, for data curators and stewards

5.2.1   Definitions of the Data Scientist

There is no well-established definition of the data scientist due to a variety of competences and skills expected from these specialists. The proposed data scientist definition is based on the definition provided in the NIST SP1500-1 document [3] and extended with the expected ability to deliver value to the organisation or to the project: *"A Data Scientist is a practitioner who has sufficient knowledge in the overlapping regimes of expertise in business needs, domain knowledge, analytical skills, and programming and systems engineering expertise to manage the end-to-end scientific method process through each stage in the big data lifecycle, till the delivery of an expected scientific and business value to science or industry."*

The NIST document defines the following groups of skills required of the data scientists: *domain experience, statistics and data mining* and *engineering* skills. The EDSF has proposed structured definition of the data scientist via definition of the related competences, skills, knowledge and proficiency level.

Initial attempt to define the data scientist has been made also by O'Reilly Strata Survey [17], seen in Chap. 2, which recognised creativity as an important feature of data scientist.

Other definitions [51] admit such desirable features as the ability to solve a variety of business problems, optimise performance and suggest new services for the organisation employing data scientist. Many practitioners admit a need for a successful data scientist to develop a special mindset, to be statistically minded, understand raw data and "appreciate data as a first class product" [52].

The qualified data scientist should be capable of working in different roles in different projects and organisations such as data engineer, data analyst or data architect, data steward, etc., and possess the necessary skills to effectively operate components of the complex data infrastructure and processing applications through all stages of the data life cycle and be able to deliver expected scientific and business values to science and/or industry.

The Data Science Competence Framework defined the main competence groups that must be possessed by data science practitioners to be able to work at different roles in the data-driven organisations:

- Data science analytics (including statistical analysis, machine learning, data mining, business analytics, others)
- Data science engineering (including software and applications engineering, data warehousing, big data infrastructure and tools)
- Domain knowledge and expertise (subject/scientific domain related)
- Data management and governance (including data stewardship, curation and preservation)
- Research methods for research-related professions and business process management for business-related professions

Detailed definition of the CF-DS competences, skills and knowledge is provided in Chap. 2.

5.2.2 Taxonomy of Data Science Occupations by Extending ESCO Hierarchy

The presented initial taxonomy of Data Science Professional profiles/roles is based on the ESCO occupations classification where proposed profiles' competences and organisational roles are defined similar to CWA 16458 ICT profiles.

The proposed new occupations do not include variety of the data science and analytics-enabled professions in different industry and research domains and sectors, which are becoming popular and highly demanded by modern organisations implementing data-driven business model.

The following suggestions were used when constructing the proposed taxonomy:

- Data scientist occupations depending on organisational role can be placed in the following top level hierarchies:

  - Managers (for managerial roles)
  - Professionals (for analytics applications developers and for infrastructure and data centre engineers); technicians and associate professionals (for operators and technicians)

- Correspondingly, new third-level occupation groups are proposed:

  - Data science/big data infrastructure managers
  - Data science professionals
  - Data science technology professionals
  - Data and information entry and access

- Group of occupations related to digital librarians, data archives management, data curations and support currently placed in the third group "*Professionals > Information and communications technology professionals > Data Science technology professionals > Data handling professionals not elsewhere classified*"; however, potentially it can also put in a new second-level group "*Clerical support workers > Data handling support workers (alternative)*". Motivation for this is the growing need for data support workers in all domains of human activities in the digital data-driven economy.
- It is recognised that existing ESCO group *database and network professionals* should be extended with new occupations (or professions) related to big data and scientific data-related profiles, examples of which are included in the table: large-scale (cloud) database administrator/operator and scientific database administrator/operator; however, further identification of such occupations needs to be done.

In the following, the data science-related occupations extension to ESCO classification is presented for the four top classification of occupations: managers, professionals, technicians and associate professionals and clerical support workers. The occupations are presented from the top level (TL) to the occupations (O) with the existing (EH) and new (NH) hierarchies for each top level, and the occupations group (OG):

- **Managers (TL)**

  - Production and specialised services managers (EH)
  - Data science/big data infrastructure managers (NH)

    Research infrastructure managers (OG)

    DSP01. Data science (group) manager (O)
    DSP02. Data science infrastructure manager (O)
    DSP03. Research infrastructure manager (O)

- **Professionals (TL)**

  - Science and engineering professionals (EH)
  - Data science professionals (NH)

    Data science professionals not elsewhere classified (OG)

    DSP04. Data scientist (O)
    DSP05. Data science researcher (O)
    DSP06. Data science architect (O)
    DSP07. Data science (application) programmer/engineer (O)
    DSP08. (Big) data analyst (O)
    DSP09. Business analyst (O)

  - Information and communications technology professionals (EH)
  - Data science technology professionals (NH)

    Data handling professionals not elsewhere classified (OG)

    DSP10. Data steward (O)
    DSP11. Digital data curator (O)
    DSP12. Data librarian (O)
    DSP13. Data archivist (O)

  - Science and engineering professionals (EH)
  - Database and network professionals (NH)

    Large-scale (cloud) data storage designers and administrators (OG)

    DSP14. Large-scale (cloud) database designer[1] (O)

    Large-scale (cloud) data storage designers and administrators (OG)

    DSP15. Large-scale (cloud) database administrator (O) (see footnote 1)

    Database and network professionals not elsewhere classified (OG)

    DSP16. Scientific database administrator (O) (see footnote 1)

---

[1]The proposed new occupations do not include variety of the data science and analytics-enabled professions in different industry and research domains and sectors.

- **Technicians and associate professionals (TL)**
  - – Science and engineering associate professionals (EH)
  - – Data science technology professionals (NH)

    Data infrastructure engineers and technicians (OG)

    DSP17. Big data facilities operators (O)
    DSP18. Large-scale (cloud) data storage operators (O)

    Database and network professionals not elsewhere classified (OG)

    DSP19. Scientific database operator (O) (see footnote 1)

- **Clerical support workers (TL)**
  - – General and keyboard clerks (EH)
  - – Data handling and support workers (NH)

    Data and information entry and access (OG)

    DSP20. Data entry/access desk/terminal workers (O)
    DSP21. Data entry field workers (O)
    DSP22. User support data services (O)

The following is the commonly used definition of the digital librarian responsibilities and functions: selection, acquisition, organisation, accessibility and preservation of digital information/library. Manages digital materials; takes a lead role in the creation, maintenance and stewardship of digital collections, including the digitisation of special collections; and develops strategies for effective management and preservation of library digital assets.

5.2.3 Definition of the Data Science Professional Profiles

From the classification presented in the previous section, this section provides definition of the Data Science Professional Profiles by defining their competences and organisational roles. The proposed definition can be instrumental in defining education and training profiles for students and for practitioners to acquire necessary competences and knowledge for specific professional profiles or occupations. It can be also used for defining certification profiles or career path building.

The presented DSPP together with CF-DS and other EDSF documents that provide the basis for multiple practical uses include but not limited to:

- Assessment of individual and team competences, as well as balanced data science team composition comprising of the data science-related roles that altogether provide necessary set of skills

- Developing tailored curriculum for academic education or professional training, in particular to bridge skills gap and staff up/re-skilling
- Professional certification and self-training

As have been said, the data science occupation groups are placed in the following top level ESCO hierarchies: managers; professionals; technicians and associate professionals; and optionally, some data management occupations can be also placed into the clerical support workers group such as digital data archivist and digital librarians.

Correspondingly, the following new third-level occupation groups are proposed:

- Data science/big data infrastructure managers
- Data science professionals
- Data science technology professionals
- Data and information entry and access (this is a candidate group under clerical support workers top level hierarchy)

It is proposed that the existing ESCO group "database and network professionals" should be extended with new occupations (or professions) related to big data or cloud-based databases: large-scale (cloud) database administrator/operator and scientific database administrator/operator; however, further identification of such occupations needs to be done.

A group of occupations related to digital librarians, data archives management, data stewardship and data curation are currently placed in the third proposed group: professionals > information and communications technology professionals > data science technology professionals > data handling professionals not elsewhere classified. However potentially it can also be added in a new second-level group *"Clerical support workers > Data handling support workers (alternative)"*. The motivation for this is a growing need for data support workers in all domains of human activities in the digital data-driven economy.

To ensure a smooth data science professions acceptance by industry and employment bodies, the proposed profiles should be compatible with the relevant standards ESCO, CWA 16458 2012 ICT Profiles, eCFv3.0 (future CEN standard EN 16324) .

Figure 5.4 graphically illustrates the existing ESCO hierarchy and the proposed new data science classification groups and corresponding new data science-related profiles. The table in the figure illustrates what competence groups are relevant to each profile by indicating competence relevance from 0 to 5 (0—not relevant, 5— very important) where information is taken from Table 4.3 that will be presented later. Figure 5.5 provides visual presentation of the identified DSPP and their grouping by the proposed high-level classification groups.

In the following, a definition of the profile summary statement of the Data Science Professional Profiles defined in the taxonomy of 5.2.2. is presented (between brackets the alternative titles and legacy titles):

- **Managers**

  – Data science/big data infrastructure managers

     Research infrastructure managers

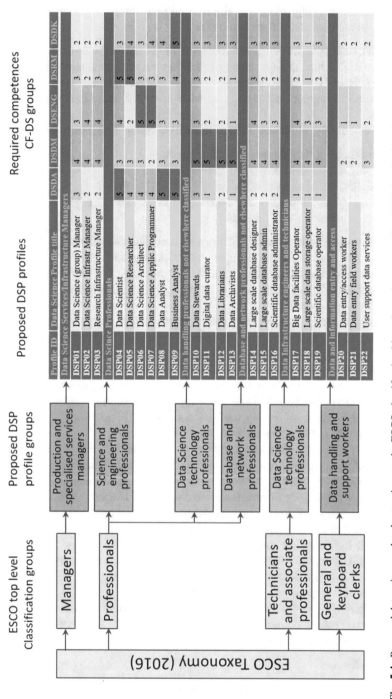

Profile ID	Data Science Profile title	DSDA	DSDM	DSENG	DSRM	DSDK
Data Science Services/Infrastructure Managers						
DSP01	Data Science (group) Manager	3	4	3	3	3
DSP02	Data Science Infrastr Manager	2	4	4	2	2
DSP03	Research Infrastructure Manager	2	4	4	3	2
Data Science Professionals						
DSP04	Data Scientist	5	3	4	5	3
DSP05	Data Science Researcher	4	3	2	5	4
DSP06	Data Science Architect	4	3	5	3	3
DSP07	Data Science Applic Programmer	4	2	5	3	4
DSP08	Data Analyst	5	3	3	3	4
DSP09	Business Analyst	5	3	3	4	5
Data handling professionals not elsewhere classified						
DSP10	Data Stewards	3	5	3	3	3
DSP11	Digital data curator	1	5	2	2	3
DSP12	Data Librarians	2	5	2	3	3
DSP13	Data Archivists	1	5	1	1	3
Database and network professionals not elsewhere classified						
DSP14	Large scale database designer	2	4	4	3	3
DSP15	Large scale database admin	2	3	3	2	3
DSP16	Scientific database administrator	2	4	3	2	3
Data Infrastructure engineers and technicians						
DSP17	Big Data facilities Operator	1	4	4	2	3
DSP18	Large scale data storage operator	1	4	3	1	1
DSP19	Scientific database operator	1	4	3	2	3
Data and information entry and access						
DSP20	Data entry/access worker		2	1		2
DSP21	Data entry field workers		2	1		2
DSP22	User support data services	3	2			2

Fig. 5.4 Proposed data science-related extensions to the ESCO classification hierarchy and corresponding DSPP by classification groups

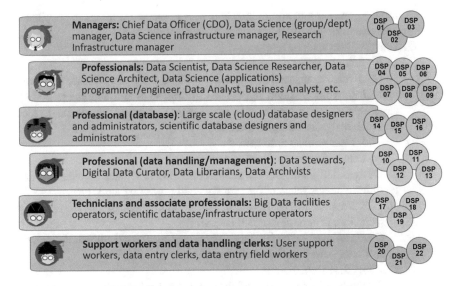

Fig. 5.5 Data Science Professional Profiles and their grouping by the proposed new professional groups compliant with the ESCO taxonomy

1. DSP01. Data science (group) manager
   (data analytics department manager)
   Proposes, plans and manages functional and technical evolutions of the data science operations within the relevant domain (technical, research, business)
2. DSP02. Data science infrastructure manager
   (research infrastructure data storage facilities manager)
   Proposes, plans and manages functional and technical evolutions of the big data infrastructure within the relevant domain (technical, research, business).
3. DSP03. Research infrastructure manager
   (research infrastructure data storage facilities manager)
   Proposes, plans and manages functional and technical evolutions of the research infrastructure within the relevant scientific domain.

- **Professionals**

  – Data science professionals

    Data science professionals not elsewhere classified

4. DSP04. Data scientist
   (data analyst)
   Data scientists find and interpret rich data sources, manage large amounts of data, merge data sources, ensure consistency of datasets and create visualisations to aid in understanding data. Build mathematical

models, present and communicate data insights and findings to specialists
and scientists and recommend ways to apply the data.
5. DSP05. Data science researcher
    (data analyst)
    Data science researcher applies scientific discovery research/process,
    including hypothesis and hypothesis testing, to obtain actionable knowl-
    edge related to scientific problem, business process, or reveal hidden
    relations between multiple processes.
6. DSP06. Data science architect
    (system architect, applications architect)
    Designs and maintains the architecture of data science applications and
    facilities. Creates relevant data models and processes workflows.
7. DSP07. Data science (application) programmer/engineer
    (scientific programmer, data engineer)
    Designs/develops/codes large data analytics applications to support
    scientific or enterprise/business processes.
8. DSP08. (Big) Data analyst
    Analyses a large variety of data to extract information about system,
    service or organisation performance and presents them in usable/
    actionable form.
9. DSP09. Business analyst
    (business development manager (data science role))
    Analyses a large variety of data Information system for improving
    business performance.

- Data science technology professionals

    Data handling professionals not elsewhere classified[2]

    10. DSP10. Data steward
        Plans, implements and manages (research) data input, storage, search,
        presentation; creates data model for domain specific data; supports and
        advises domain scientists/researchers. Creates data model for domain-
        specific data, supports and advises domain scientists/researchers during
        the whole research cycle and data management life cycle
    11. DSP11. Digital data curator
        (digital curator, digital archivist, digital librarian)
        Finds, selects, organises, shares (exhibits) digital data collections,
        maintains their integrity, up-to-date status and freshness, discoverability.
    12. DSP12. Data librarian
        (Digital data curator)
        Data librarians perform or support one or more of the following:
        acquisition (collection development), organisation (cataloguing and

---

[2]The proposed professional (data handling/management) taxonomy group does not include the
occupation of the digital librarian as primarily related to digitising the library resources.

metadata) and the implementation of appropriate user services. Data librarians apply traditional librarianship principles and practices to data management, including data citation, digital object identifiers (DOIs), ethics and metadata.

13. DSP13. Data archivist
    (digital archivists)
    Maintain historically significant collections of datasets, documents and records and other electronic data and seek out new items for archiving.

– Database and network professionals.

  Large-scale (cloud) data storage designers and administrators

14. DSP14. Large-scale (cloud) database designer
    (data engineer, data architect)
    Designs/develops/codes large-scale databases and their use in domain/ subject-specific applications according to the customer needs.
15. DSP15. Large-scale (cloud) database administrator
    Designs and implements or monitors and maintains large-scale cloud databases.
16. DSP16. Scientific database administrator (see footnote 1)
    (large-scale (cloud) database administrator)
    Designs and implements or monitors and maintains large-scale scientific databases.

• **Technicians and associate professionals**

  – Data science technology professionals

    Data infrastructure engineers and technicians

17. DSP17. Big data facilities operators
    Manages daily operation of facilities and resources and responds to customer requests. Includes all operations related to data management and data life cycle.
18. DSP18. Large-scale (cloud) data storage operators
    Manages daily operation of cloud storage, including related to data life cycle, and responds to requests from storage users
19. DSP19. Scientific database operator (see footnote 1)
    (Large-scale (cloud) data storage operators)
    Manages daily operation of scientific databases, including related to data life cycle, and responds to requests from database users.

• **Clerical support workers**

  – Data handling and support workers (NH)

    Data and information entry and access (OG)

20. DSP20. Data entry/access desk/terminal workers
    (Data entry desk/terminal worker)
    Enter data into data management systems directly reading them from source, documents or obtained from people/users
21. DSP21. Data entry field workers
    The same work done on field when collecting data from disconnected sensors or doing direct counting or reading
22. DSP22. User support data services
    User support data services. Support users to entry their data into governmental service and user facing applications.

5.2.4   Recognising New Emerging Profession in Data Management: Data Steward

Recognising importance of the data steward as a new emerging profession in a typical research institution, the DSPP provides the initial definition of the data steward professional profile: *Data steward is a data handling and management professional whose responsibilities include planning, implementing and managing (research) data input, storage, search and presentation. Data steward creates data model for domain-specific data and supports and advises domain scientists/ researchers during the whole research cycle and data management life cycle.*

The data steward organisational role was first defined in the first version of the DAMA DMBOK in 2009 and present in the current edition of DMBOK. The important role of the data steward for managing research data is recognised in the HLEG report on European Open Science Cloud (October 2016) [53] which identified the critical need for core data experts and data stewards in particular. Data steward competences definition and training is an important component of the GO FAIR initiative [54, 55] involving wide international cooperation and numerous GO FAIR national competence centres promoting FAIR data principles in research data management (findable, accessible, interoperable, reusable).

Further development in the Horizon 2020 EOSC pilot project (2016–2018) activity [56, 57] defined stewardship as a shared responsibility of professional groups involved into data management: data management and curation, data science and analytics, data services engineering and domain research [25]. Competences and skills groups and organisational roles are defined around typical processes and stages in data management: plan and design, capture and process, integrate and analyse, apprise and present, publish and release, expose and discover, govern and assess, scope and resource, advise and enable.

The newly started EU funded project FAIRsFAIR [58] targets to further promote data stewardship professionalisation by developing data stewardship curriculum and include FAIR data management principles into academic practice.

5.2.5   Mapping Data Science-Related Competences to Professional Profiles

The CF-DS competence groups are defined in Chap. 2 as follows:

1. Data analytics (DSDA). Use appropriate statistical techniques and predictive analytics on available data to deliver insights and discover new relations.
2. Data management (DSDM). Develop and implement a data management strategy for data collection, storage, preservation and availability for further processing.
3. Data science engineering (DSENG). Use engineering principles to research, design, develop and implement new instruments and applications for data collection, analysis and management
4. Research methods and project management (DSRMP) for research domain and business process management (DSBPM). Create new understandings and capabilities by using the scientific method (hypothesis, test/artefact, evaluation) or similar engineering methods to discover new approaches to create new knowledge and achieve research or organisational goals.
5. Data science domain knowledge (DSDK). Use domain knowledge (scientific or business) to develop relevant data analytics applications, and adopt general data science methods to domain-specific data types and presentations, data and process models, organisational roles and relations.

Taking into account the definitions from above and the professional profiles defined in the previous sections, Table 4.3 provides a mapping between professional profiles and data science competence groups together with the suggested ranking the relevance of different competence groups to corresponding data science profiles (where 0 is not relevant and 5 is highly relevant). Note, Table 5.2 is using scale 0 to 5 for the competence groups relevance.

5.2.6   Data Science Team Composition

Data science team composition and competences matching is one of intended uses of the EDSF and DSPP in particular. Figure 5.6 illustrates a case of creating a data science team or group for an average size of the research organisation with affiliated number of researchers 200–300, which would require a data science team of 10–15 members whose responsibility would include supporting all main stages of data life cycle: data collection, data input/ingest, data analysis, reporting, visualisation and storage. The figure also illustrates possible roles that may be assigned to perform different functions at different data workflow stages.

To support all data-related research or production stages, the following roles may be required (including suggested staffing for the team of 10–12 members):

- (Managing) Data science architect (1)
- Data scientist (1), data analyst (1)

Table 5.2 Mapping data science competence groups to the proposed profiles

	Data Science Competences Groups (relevance 1 - low, 5 – high)				
Profile ID	DSDA	DSDM	DSENG	DSRM	DSDK
DSP01	3	4	3	3	2
DSP02	2	4	4	2	2
DSP03	2	4	4	3	2
DSP04	5	3	4	5	3
DSP05	4	3	2	5	4
DSP06	4	3	5	3	3
DSP07	4	2	5	3	4
DSP08	5	3	3	3	4
DSP09	5	3	3	4	5
DSP10	3	5	3	3	3
DSP11	1	5	2	2	3
DSP12	2	5	2	2	3
DSP13	1	5	1	1	3
DSP14	2	4	4	3	3
DSP15	2	4	3	2	3
DSP16	2	4	3	2	3
DSP17	1	4	4	2	3
DSP18	1	4	3	1	1
DSP19	1	4	3	2	3
DSP20	0	2	1	0	2
DSP21	0	2	1	0	2
DSP22	0	3	2	0	2

- Data science application architect/developer/programmer (2)
- Data infrastructure/facilities administrator/operator: storage, cloud, computing (1)
- Data stewards, curators, archivists (3–5)

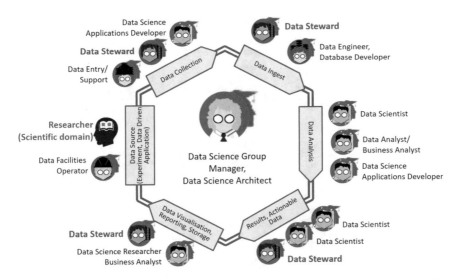

Fig. 5.6 Matching the candidate's competences for the data scientist competence profile (as defined in the DSPP document [4])

It is possible that some of the above roles can be redefined and reallocated to the data science team from the previous ICT and IT infrastructure groups or departments. In this case, some basic data science training will be required for not initially data-related professions.

It also suggested a distinct role of the data steward, a new emerging role for data-driven research organisations and projects. Data steward should play a bridging role between the subject domain researcher and the data science team or data scientist in particular cases to help to translate between subject domain and data science or data analytics domain. Data stewards can have both backgrounds either ICT and computer or digital curation/librarian.

Similar approach to data science and data governance roles definition and team building was used in IBM enterprise consulting practice [59].

5.2.7 Data Science-Enabled Professions

Recent studies by BHEF, PwC [60] and IBM, BGT and BHEF [61] identified strong growth of the data science and analytics- (DSA) enabled jobs that are not are not pure data scientists but require extensive DSA knowledge to work in the specific industry sectors. Figure 5.7 from PwC and BHEF study [16] provides illustration of currently highly demanded DSA-enabled jobs in multiple industry and business sectors: finance and insurance; healthcare and social assistance; information; manufacturing; professional, scientific and technical Services; retail trade.

The study provides data of 2.35 million job postings in the USA in 2017: 23% data scientist and 67% DSA-enabled jobs. It is also strong demand for managers and

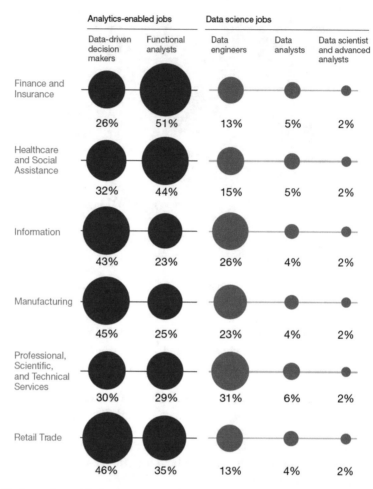

Fig. 5.7 Strong demand for business people with analytics skills, not just data scientists in multiple industry sectors [16]

decision-makers with the data science (data analytics) skills/understanding. This creates a new challenge to deliver actionable knowledge and competences to CEO level managers.

5.3   Competence, Proficiency and Qualifications Levels for Professional Profiles in Data Science

5.3.1   Mapping Between CF-DS and eCF

The CF-DS competences are defined using proposed new competence groups provide the basis for defining new competences related to data science that can be

Table 5.3 Proposed e-CF3.0 extension with the data science-related competences

Competence group	Competences related to data science	Corresponding CF-DS competence groups
A. PLAN (and design)	A.10* Organisational workflow/processes model definition/formalisation A.11* Data models and data structures	DSDA DSENG
B. BUILD (Develop and deploy/ implement)	B.7* Apply data analytics methods (to organisational processes/data) B.8* Data analytics application development B.9* Data management applications and tools B.10* Data science infrastructure deployment (including computing, storage and network facilities)	DSDA DSENG DSDM
C. RUN (operate)	C.5* User/usage data/statistics analysis C.6* Service delivery/quality data monitoring	DSDM DSENG
D. ENABLE (use/utilise)	D10. Information and knowledge management (powered by data science analytics)—*refactored* D.13* Data analysis, insight or actionable information extraction, visualisation D.14* Support business processes/roles with data analytics, visualisation and reporting (support to D.5, D.6, D.7, D.12) D.15* Data management, curation, preservation, provenance	DSDA DSDK/DSBA
E. MANAGE	E.10* Support management and business improvement with data and insight (data-driven organisational processes management) (support to E.5, E.6) E.11* Data analytics for (business) Risk analysis/management (support to E.3) E.12* ICT and information security monitoring and analysis (support to E.8)	DSDA DSENG DSDM

added to the existing e-CF3.0. In particular, the EDISON Project suggested the following additional e-competences related to data scientist functions as listed in Table 5.3 (assigned numbers are continuation of the current e-CF3.0 numbering). When defining individual professional profile or role, the presented data science competences can be combined with the generic competences listed in the original e-CF3.0 reflecting that beneficially the data scientist needs to have basic or advanced knowledge and skills in general ICT domain. The e-CFv3.0 (2018) has limited data and data science-related content and will benefit from the perspective of the data science competences framework. However, the broader scope and granularity of the e-CF will determine the transferability of these suggestions in part or entirety.

The EN 16234-1 *e-Competence Framework* (almost referred to as eCFv4.0) is a European standard developed by CEN TC 428 ICT Professionalism and Digital Competences Working Group. New revision of EN 16234-1 includes the following new competences roles and skills related to data science and defined in correspondence with CF-DS:

D7. Data Science and Analysis Uses and applies data analytic techniques such as data mining, machine learning, prescriptive and predictive analytics to apply data insight to address organisation's challenges and opportunities.

Data Specialist Role Endures the implementation of the organisations data management policy.

Data Scientist Leads the process of applying data analytics. Delivers insights from data by optimising the analytics process and presenting visual data representations. Skill K11 FAIR data management principles (findability, accessibility, interoperability, reusability) and Skill K6 data governance, data governance strategy, data management plan (DMP).

5.3.2 Mapping Between CF-DS, e-CF Proficiency Levels and EQF Qualification Levels

Both EN 16234-1 and e-CF3.0 defines 5 proficiency levels applied to individual competences; however, individual competence may not necessarily have all proficiency levels but can require lower or upper set of proficiency demanding on the complexity of performed tasks. The following table provides a guide to how this might be expressed. For the complete table go to Annex 2 of the e-CF 3.0.

Establishing a "level" relationship between CF-DS competences and that of the e-CF will support orientation and provide a general context for the positioning of job roles constructed from competence combinations. Although direct connections between the e-CF and EQF are not possible, as they represent different concepts, expressing a general relationship helps to clarify the complexity of tasks or activities incorporated within a competence.

Table 5.4 offers a potential connection between data science competence, e-CF competence and EQF qualification levels.

Table 5.5 in the next pages provides details on the e-CF proficiency levels, e-CFL and corresponding definition of the EQF levels (EQFL)

Table 5.4 Mapping between CF-DS, e-CF and EQF levels

CF-DS competence proficiency levels	e-CF proficiency levels	EQF qualification levels
Associate/entry	e-1	eqf-3, eqf-4
Professional	e-2, e-3	eqf-5, eqf-6
Expert/lead	e-4, e-5	eqf-7, eqf-8

Table 5.5 Mapping between e-CF proficiency levels and EQF qualification levels

e-CFL	e-CF levels description	EQFL	EQF levels description
e-5	**Principal** Overall accountability and responsibility; recognised inside and outside the organisation for innovative solutions and for shaping the future using outstanding leading-edge thinking and knowledge	eqf8	Knowledge at the most advanced frontier, the most advanced and specialised skills and techniques to solve critical problems in research and/or innovation, demonstrating substantial authority, innovation, autonomy, scholarly or professional integrity
e-4	**Lead professional/senior manager** Extensive scope of responsibilities deploying specialised integration capability in complex environments; full responsibility for strategic development of staff working in unfamiliar and unpredictable situations	eqf7	Highly specialised knowledge, some of which is at the forefront of knowledge in a field of work or study, as the basis for original thinking, critical awareness of knowledge issues in a field and at the interface between different fields, specialised problem-solving skills in research and/or innovation to develop new knowledge and procedures and to integrate knowledge from different fields, managing and transforming work or study contexts that are complex, unpredictable and require new strategic approaches, taking responsibility for contributing to professional knowledge and practice and/or for reviewing the strategic performance of teams
e-3	**Senior professional/manager** Respected for innovative methods and use of initiative in specific technical or business areas; providing leadership and taking responsibility for team performances and development in unpredictable environments	eqf6	Advanced knowledge of a field of work or study, involving a critical understanding of theories and principles, advanced skills, demonstrating mastery and innovation in solving complex and unpredictable problems in a specialised field of work or study, management of complex technical or professional activities or projects, taking responsibility for decision-making in unpredictable work or study contexts, for continuing personal and group professional development
e-2	**Professional** Operates with capability and independence in specified boundaries and may supervise others in this environment; conceptual and abstract model building using creative thinking; uses theoretical knowledge and practical skills to solve complex problems within a predictable and sometimes unpredictable context	eqf5	Comprehensive, specialised, factual and theoretical knowledge within a field of work or study and an awareness of the boundaries of that knowledge, expertise in a comprehensive range of cognitive and practical skills in developing creative solutions to abstract problems, management and supervision in contexts where there is unpredictable change, reviewing and developing performance of self and others.

(continued)

Table 5.5 (continued)

e-CFL	e-CF levels description	EQFL	EQF levels description
		eqf4	Factual and theoretical knowledge in broad contexts within a field of work or study, expertise in a range of cognitive and practical skills in generating solutions to specific problems in a field of work or study, self-management within the guidelines of work or study contexts that are usually predictable, but are subject to change, supervising the routine work of others, taking some responsibility for the evaluation and improvement of work or study activities
e-1	**Associate** Able to apply knowledge and skills to solve straightforward problems; responsible for own actions; operating in a stable environment	eqf3	Knowledge of facts, principles, processes and general concepts, in a field of work or study, a range of cognitive and practical skills in accomplishing tasks. Problem-solving with basic methods, tools, materials and information, responsibility for completion of tasks in work or study, adapting their own behaviour to circumstances in solving problems

Chapter 6
Use Cases and Applications

Yuri Demchenko, Luca Comminiello, Tomasz Wiktorski, Juan J. Cuadrado-Gallego, Oleg Chertov, Ernestina Menasalvas, Ana M. Moreno, Nik Swoboda, and Steve Brewer

This chapter includes a set of use cases and examples of practical use of EDSF for developing data science curricula, competences assessment, data science team building and addressing new skills demand for emerging data economy.

The presented applications and use cases have been written by different authors and are presented completely with its own references at the end of the section; for that reason, these sections can be considered as sub-chapters. All of them strongly rely and refer to the EDSF components definition in Chaps. 1–5.

This has four sections; in each one of them a different use case or application of the EDSF is presented:

Y. Demchenko
Universiteit van Amsterdam, Amsterdam, The Netherlands
e-mail: y.demchenko@uva.nl

L. Comminiello
Universidad degli Studi de Perugia, Perugia, Italy

T. Wiktorski
Universitetet i Stavanger, Stavanger, Norway
e-mail: tomasz.wiktorski@uis.no

J. J. Cuadrado-Gallego (✉)
Department of Computer Science, University of Alcalá, Madrid, Spain
e-mail: jjcg@uah.es

O. Chertov
Igor Sikorsky Kyiv Polytechnic Institute, National Technical University of Ukraine, Kiev, Ukraine

E. Menasalvas · A. M. Moreno · N. Swoboda
Universidad Politécnica de Madrid, Madrid, Spain
e-mail: emenasalvas@fi.upm.es; ammoreno@fi.upm.es; nswoboda@fi.upm.es

S. Brewer
University of Southampton, Southampton, UK
e-mail: SBrewer@lincoln.ac.uk

© Springer Nature Switzerland AG 2020
J. J. Cuadrado-Gallego, Y. Demchenko (eds.), *The Data Science Framework*,
https://doi.org/10.1007/978-3-030-51023-7_6

- Section 6.1, by Yuri Demchenko, from the Universiteit van Amsterdam, Amsterdam, the Netherlands, and Luca Comminiello, from the Universidad degli Studi de Perugia, Italy, presents "Designing Customisable Data Science Curriculum Using Ontology for Data Science Competences and Body of Knowledge". If the reader is interested to expand the information in this section papers [62, 63] can be read.
- Section 6.2, by Yuri Demchenko; Tomasz Wiktorski, from the Universitetet i Stavanger, Stavenger, Norway; Juan J. Cuadrado-Gallego, from the Univesidad de Alcalá, Madrid, Spain; and Oleg Chertov, from the Igor Sikorsky Kyiv Polytechnic Institute of the National Technical University of Ukraine, presents "Analysis of Big Data Platforms and Tools for Data Analytics in the Big Data and Data Science Curricula". If the reader is interested to expand the information in this section, papers [64, 65] can be read.
- Section 6.3, by Ernestina Menasalvas, Ana M. Moreno and Nik Swoboda, from the Universidad Politécnica de Madrid, Spain, presents the "Definition and use of Big Data Value Data Science Badges".
- Section 6.4, by Yuri Demchenko and Steve Brewer, the University of Southampton, Southampton, UK, presents the figure of "Data Management and Data Stewardship for Industry, Research and Academia". If the reader is interested to expand the information in this section, papers [66, 67] can be read.

6.1   Designing Customisable Data Science Curriculum Using Ontology for Data Science Competences and Body of Knowledge

Yuri Demchenko and Luca Comminiello

6.1.1   Introduction

Importance of data science education and training is growing with the emergence of data-driven technologies and organisational culture that intend to derive actionable value for improving research process or enterprise business using a variety of enterprise data and widely available open and social media data. Modern data-driven research and industry require new types of specialists that are capable of supporting all stages of the data life cycle from data production and input to data processing and actionable results delivery, visualisation and reporting, which can be jointly defined as the data science professions family. The education and training of data scientists requires multidisciplinary approach combining wide view of the data science and analytics foundation with deep practical knowledge in domain-specific areas. In modern conditions with the fast technology change and strong skills demand, the data science education and training should be customisable and delivered in multiple forms, also providing enough data lab facilities for practical training. This section discusses approach to building customisable data science curriculum for different types of learners based on using the ontology of the EDISON Data Science Framework (EDSF) developed in the EU-funded Project EDISON and widely used by universities and professional training organisations.

EDSF was developed with the view of multiple practical uses for the whole range of tasks faced by universities, professional training organisations, companies and certification bodies related to data science education, training and capacity management. The following are the intended practical applications of EDSF:

- Academic curriculum design for general data science education and individual learning path construction for customisable training and career development
- Professional competence benchmarking, including CV or organisational profiles matching
- Professional certification of data science professionals
- Vacancy construction tool for job advertisement (for HR) using controlled vocabulary and data science taxonomy
- Data science team building and organisational roles specification

The EDSF toolkit has being developed to support above-mentioned applications and ensure their compatibility. It contains a number of API, ontologies and datasets representing different components of the EDSF and mapping between them. EDSF toolkit is an ongoing development and available as open source at the EDSF GitHub project [6].

6.1.1.1   Demand for Data Science Competences and Customisable Curriculum

Sustainable development of the modern data-driven economy requires new type of data-driven and data science and analytics-enabled competences and workplace skills. Fast technology change and new skills demand requires rethinking and redesigning both traditional educational models and existing courses to reflect multidisciplinary nature of data science and its application domains. At present, most of the existing university curricula and training programmes cover a limited set of competences and knowledge areas of what is required for multiple data science and general data management professional profiles and organisational roles enacted by research and industry. In conditions of continuous technology development and shortened technology change cycle, data science education requires the effective combination of theoretical, practical and workplace skills.

Industry digitalisation and wide use of data-driven technologies facilitate demand for data science and analytics-enabled professions, this trend is confirmed by multiple European and global market studies. The IDG report 2017 [68] provided deep analysis of the European data market and growing demand for data workers and estimated the total number of data workers to grow from 6.1 million in 2016 to 10.4 million in 2020 where the data-related skills gap is estimated as 769,000 or 9.8%

(2020). Addressing this demand and gap is becoming critical for European economy and challenge for universities.

Business Higher Education Forum (BHEF) has published in 2017 two important reports in cooperation with Price Waterhouse Coopers, IBM and Burning Glass Technologies [21, 22] that studied data science and analytics (DSA) job market in the USA and identified a number of actions to be addressed by business, higher education, government and professional organisations to address increased demand and growing gap in demand and supply of skilled DSA workforce capable of effectively working in modern data-driven economy.

Recent OECD report [69] confirms the urgent need to address data and general digital skills for all types of workforce and economy sectors. An effective professional education should provide a foundation for future continuous professional self-development and mastering new emerging technologies, which can provide a basis for the lifelong learning model adoption. Flexibility in providing education and training curricula and course is the key adopting future skills management and capacity building models.

6.1.2   Data Science Data Model and Ontology

EDSF data model represents all the complex relations between the EDSF components such as competences, knowledge, skills, professional profiles, proficiency levels and organisational roles, which exist in real-life organisations. Initial EDSF definition followed the 4 parts structure as described in Chap. 3. Initial definition of EDSF was made in the form of Excel workbooks and table which provided a good way of documenting but was difficult to use for practical applications.

6.1.2.1   EDSF Data Model

Currently, EDSF toolkit contains a number of datasets representing different components of the EDSF and mapping between them. Future EDSF development will formally define the ontologies related to the EDSF components and related dictionaries.

Figure 6.1 illustrates the relation between different datasets and ontologies comprising EDSF. The CF-DS is structured along four dimensions (similarly to European e-Competence Framework e-CFv3.0 [15]) that include (1) competence groups, (2) individual competences definition, (3) proficiency levels and (4) corresponding knowledge and skills. In this context, each individual competence includes a set of required knowledge topics and a set of skills type A and skills type B. Such CF-DS structure allows for competence-based curriculum design where competences can be defined based on the professional profile (see DSPP specification in Chap. 5 for mapping between professional profiles and competences) or target leaners group when designing a full curriculum, or based on competence

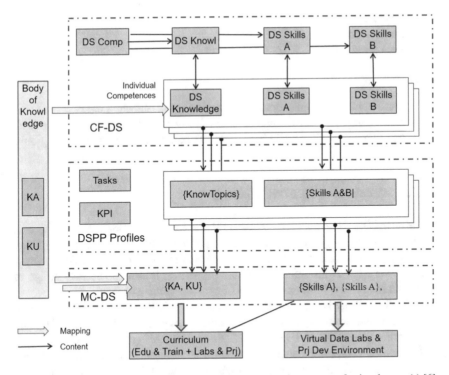

Fig. 6.1 EDSF data model and customised curriculum design for target professional group(s) [6]

benchmarking for tailored training to address identified competences and knowledge gaps.

When a set of required competences is defined together with the required ranking or proficiency level, the set of required knowledge topics can be extracted and ordered according to proficiency level and relevance (or benchmark score) for further mapping to DS-BoK knowledge areas and knowledge units. The set of KAs and KUs defined for a specific competence set defines the structure of the curriculum that further can be mapped to model curriculum learning units defined as individual courses and KAG-related course groups.

At the same time, required proficiency level is scored for each KA and KU, which will define a mastery level and corresponding learning outcome for the targeted education or training. The following mastery levels are defined (using workplace terminology that can be easy mapped to mastery levels defined in MC-DS):

- A—Awareness

  1. Understand terminology
  2. Understand principles
  3. Apply principles
  4. Understand methods

- U—Use/Application

5. Apply basics
6. Supervised use
7. Unsupervised use

- P—Professional/Expert

8. Development of applications using a wide range of technologies
9. Supervise project development, team of professionals, where borderline mastery levels 4 and 7 actually belong to both higher level and lower level groups

6.1.2.2   Definition of the EDSF Ontology

In the new EDSF Release 4 (EDSF2020) [6], the CF-DS and DS-BoK are expressed in the form of ontology that is linked also to DSPP profiles definition. The ontology provides an effective format for representing rich relations between EDSF components in the form of instance, classes and properties; it also allows easy design of APIs and benefitting from existing APIs (e.g. for Python and Java).

CF-DS ontology is a core ontology linking all EDSF entities, classes and properties. It includes ontologies for all individual competences defined for the main competence groups DSDA, DSENG, DSDM and DSRMP (refer to Chap. 3) defined as subclasses. Each competence is represented as an instance of the class to which it belongs (e.g. DSDA01 is an instance of DSDA subclass). Each competence instance includes the following properties:

- Knowledge that is required for competences, defined as knowledge topics and linked to knowledge units (KUs) in the DS-BoK
- Skills related to the knowledge topics (defined in CF-DS as Skills type A)
- Skills related to practical experience including programming, tools and platforms (defined in CF-DS as Skills type B)

Figure 6.2 illustrates the relation between different datasets and ontologies comprising EDSF; in particular it illustrates example of the DSDA01 competence that is defined as "Effectively use variety of data analytics techniques, such as Machine Learning, Data Mining, Prescriptive and Predictive Analytics, for complex data analysis through the whole data lifecycle". The DSDA01 properties include knowledge topics KDSDA*, Skills Group A SDSDA* and Skills Group B SDSA*.

The Protégé ontology editor was used for ontology design and management. It allows creating and managing an ontology through an intuitive graphic interface and permits to export the ontology in a large number of formats [70]. In this project, RDF/OWL format is chosen in order to query the ontology using the Python module, OwlReady2.

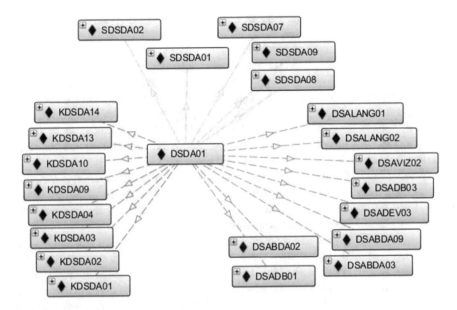

Fig. 6.2 Example DSDA01 competence and its properties

6.1.3   Data Science Curriculum Design Using Ontology

This section describes the workflow of using EDSF for curriculum design for selected/intended set of competences that are required for (1) a specific data science professional profile defined based on DSPP document or (2) individual training programme defined based on competence assessment and identified gaps. The individual competence assessment can be done based on CV matching against intended job position or professional profile. It can be also done based on the certification exam or just self-assessment questionnaire. Outcome of this process is either level of matching or competence gap that can be used for suggesting necessary training programme or tailored curriculum. As a part of the EDSF toolkit development, the authors have tested different methods for CV and job vacancy/profile matching using Doc2Vec document embedding and PV-DBOW training algorithms (available in the gensim Python libraries) [71–73].

When a set of required competences is defined together with the required ranking or proficiency level, the set of required knowledge topics can be extracted from individual competences (note, there exist multiple links from competence instances to single knowledge topic) and ordered according to proficiency level and relevance (or benchmark score) for further mapping to DS-BoK knowledge areas and knowledge units. The set of KAs and KUs defined for a specific competence set defines the structure of the curriculum that further can be mapped to the model curriculum learning units defined as individual courses and KAG-related courses groups;

otherwise, it can be used directly as advice for constructing curriculum by the programme or course manager.

At the same time, required proficiency level is scored for each KA and KU, which will define mastery levels and corresponding learning outcome for the targeted education or training. When using MC-DS as a template for designing customised curriculum, the proficiency levels (using scale 0 to 9) can be easily mapped to 3 mastery levels defined in MC-DS): familiarity, usage, assessment (refer to MC-DS specification in Chap. 4). Collected Skills type B linked to intended competences will provide advice on the required hands-on training and practical project development tasks and development platform.

When using EDSF ontology, it is a routine task to extract all required knowledge topics, map them to KA/KU and define relevance score by querying ontology with a few lines of code using OwlReady2 Python module that allows manipulating ontology classes, instances and properties transparently.

Figure 6.3 illustrates example of relations between EDSF components when extracting required knowledge units for DSDA group of competences for DSP04—Data Scientist Professional Profile (refer to DSPP specification for details). It shows that the following competences are required with the corresponding relevance/weight: DSDA01 = 9; DSD02 = 9; DSDA04 = 7. Required knowledge units are defined through the mapping knowledge topics KDSDA* to KU (using DS-BoK) and weighted based on average relevance by competences.

The same process is applied to other competence groups relevant to specific professional profile or competence gap. Figure 6.4a, b shows example of the suggested curriculum structure for two professional profiles: DSP04—data scientist and DSP10—data steward. The diagrams reflect relative structure of the curriculum where data scientist has major part of the data analytics courses (DSDA—blue) followed by necessary knowledge in data management (DSDM—orange), and data steward curriculum must focus on the data management courses (DSDM—orange), followed by basic knowledge in data analytics (DSDA—blue).

The EDSF toolkit and its outcome provides advice on the suggested curriculum structure that can be adjusted to real condition of the teaching or training institution depending on the available teaching staff and lab base. It is also important that the courses are correctly ordered, and necessary prerequisite knowledge is specified. When using third-party educational platform providers and cloud-based data labs, the presented application can provide a specification for required educational platform.

6.1.4 Cloud-Based Data Science Education Environment and Virtual Data Labs: IDE, Tools and Datasets

As outcome of the curriculum design, the application may provide suggestions that the set of Skills type A will define learning outcomes and Skills type B will provide

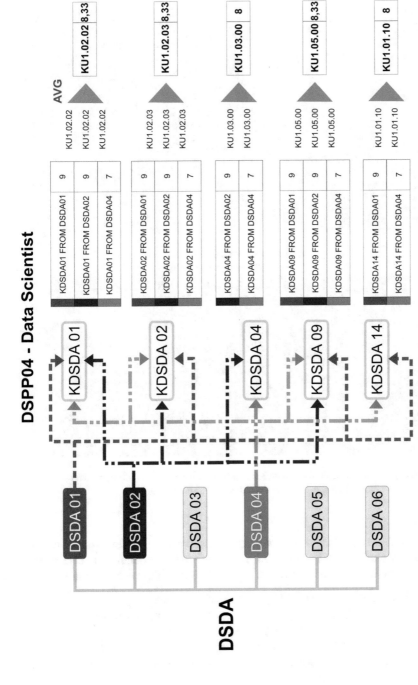

Fig. 6.3 Extracting required knowledge units from EDSF ontology

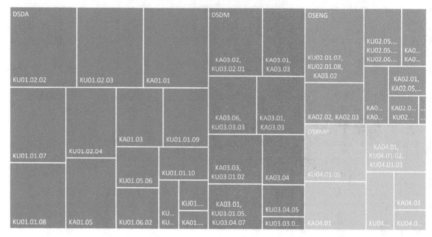

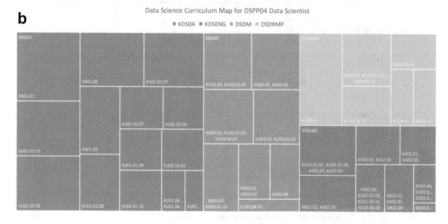

Fig. 6.4 Example curriculum structure for DSP04—data scientist and DSP10—data steward. (**a**) Data scientist curriculum structure, (**b**) Data steward curriculum structure

advice on the required hands-on training and practical project development environment and platform (refer to CF-DS document). As an example, the data scientist curriculum should include the following elements to achieve necessary skills type B:

- Python (or R) and corresponding data analytics libraries
- NoSQL and SQL Databases (Hbase, MongoDB, Cassandra, Redis, Accumulo, MS SQL, My SQL, PostgreSQL, etc.)
- Big data analytics platforms (Hadoop, Spark, Data Lakes, others)
- Real-time and streaming analytics systems (Flume, Kafka, Storm)

- Kaggle competition, resources and community platform
- Visualisation software (D3.js, Processing, Tableau, Julia, Raphael, etc.)
- API management and web scraping
- Git versioning system as a general platform for software development
- Development frameworks: Python, Java or C/C++, AJAX (asynchronous JavaScript and XML), D3.js (data-driven documents), jQuery, others
- Cloud-based big data and data analytics platforms and services, including large-scale storage systems.

Using cloud resources to build effective and up-to-date professional data science education environment is inevitable with current fast technology development and required computational performance that can be requested on-demand.

Major cloud service providers (CSP) provide a wide range of data analytics and business analytics services and platforms that can be equally used by big, small and medium companies and individuals on a pay-per-use basis. In addition to the possibility of using the same resources for education and training purposes, the major CSPs provide designated education and self-training resources that are in many cases supported also by educational grants for students and teachers.

The following cloud-based resources from the major cloud providers can be used to build hybrid Data Science Education Environment (DSEE) and VDLabs (in addition to regular compute and storage resources):

- Microsoft Azure Data Lakes Analytics, Power BI, HDInsight Hadoop-as-a-Service, others
- AWS Elastic MapReduce (EMR), QuickSight, Kinesis and wide collection of open datasets
- IBM Data Science Experience, Data Labs, Watson Analytics

An important component of data science education is educational datasets that often need to be provided with their specific applications. While many educational datasets are available from above-mentioned cloud platforms, from community run Kaggle [74] and UCI Machine Learning Repository [75], use of cloud-based VDLabs allows us to instantiate the whole experimental set-up or environment together with used datasets in the case of specific domain-focused education or training.

6.1.5 Conclusions

EDSF provides a common semantic basis for interoperability of all forms of the data science curriculum definition and education or training delivery, as well as knowledge assessment based on fully enumerated definition of EDSF components and individual units. Besides defining academic components of the effective and consistent curriculum, EDSF also provides advice on the required Data Science Education Environment to facilitate fast practical knowledge and skills acquisition by students and learners.

Further EDSF toolkit development will include defining ontologies for MC-DS and DSPP that are intended to be compatible with the ESCO ontologies [7] that are defined as a European standard for competences, skills and occupations definition.

The EDSF and its proposed further integration with the Data Science Education Environment will facilitate education and training for highly demanded data science and analytics competences and skills.

The EDSF maintenance and continuous development as well as collection of the best practices in data science education and training is supported and coordinated by the EDISON community, in cooperation with national and EU projects as well as supported by the Research Data Alliance (RDA) Interest Group on Education and Training on Handling Research Data (IG-ETHRD) [4]. Participation and contribution to both the IG-ETHRD and EDSF Community Initiative is open and free.

6.2   Big Data Platforms and Tools for Data Analytics in the Big Data and Data Science Curricula

Yuri Demchenko, Tomasz Wiktorski, Juan J. Cuadrado-Gallego, and Oleg Chertov

6.2.1   Introduction

Modern data science and business analytics applications extensively use big data infrastructure technologies and tools which are commonly cloud based and are available at all major cloud platforms. Knowledge and ability to work with the modern big data platforms and tools to effectively develop and operate the data analytics applications are required of the modern data science practitioners. Including big data infrastructure topics into the general data science curriculum will help the graduates to easily integrate into the future workplace.

This paper refers to and effectively uses the EDISON Data Science Framework (EDSF), initially developed in the EDISON Project (2015–2017) and currently maintained by the EDISON community [6]. The EDSF provides a general framework for the data science education, curriculum design and competences management which has been discussed in the author's previous works [63, 76]. Big Data Infrastructure Technologies (BDIT) is a part of the defined in EDSF the Data Science Engineering Body of Knowledge (DSENG-BoK) and Model Curriculum (MC-DSENG) described in detail below.

This paper is focused on the definition of the Data Science Engineering Body of Knowledge and Big Data Infrastructure Technologies for Data Analytics (BDIT4DA) course. The paper provides a brief overview of the big data infrastructure technologies and existing cloud-based platforms and tools for big data processing and data analytics which are relevant to the BDIT4DA course. The focus is given on the cloud-based big data infrastructure and analytics solutions and in particular to understanding and using the Apache Hadoop ecosystem as the major big data platform, its main functional components MapReduce, Spark, HBase, Hive, Pig and supported programming languages Pig Latin and HiveQL.

Knowledge and basic experience with the major cloud service providers (e.g. Amazon Web Services (AWS), Microsoft Azure and Google Cloud Platform GCP) as well as the Cloudera Hadoop Cluster or Hortonworks Data Platform are important to develop necessary knowledge and strong practical skills. These topics need to be included in both lecture course and hands-on practice.

6.2.2 *Data Science Engineering Body of Knowledge*

Data science engineering knowledge group builds the ability to use engineering principles to research, design, develop and implement new instruments and applications for data collection, analysis and management. It includes knowledge areas that cover software and infrastructure engineering, manipulating and analysing complex, high-volume, high-dimensionality data, structured and unstructured data, cloud-based data storage and data management.

Data science engineering includes software development, infrastructure operations and algorithms design with the goal to support big data and data science applications in and outside the cloud. The following are commonly defined data science engineering knowledge areas (as part of KAG02-DSENG):

- KA02.01 (DSENG/BDI) Big data infrastructure and technologies, including NOSQL databases, platforms for big data deployment and technologies for large-scale storage
- KA02.02 (DSENG/DSIAPP) Infrastructure and platforms for data science applications, including typical frameworks such as Spark and Hadoop, data processing models and consideration of common data inputs at scale
- KA02.03 (DSENG/CCT) Cloud computing technologies for big data and data analytics
- KA02.04 (DSENG/SEC) Data and applications security, accountability, certification and compliance
- KA02.05 (DSENG/BDSE) Big data systems organisation and engineering, including approaches to big data analysis and common MapReduce algorithms
- KA02.06 (DSENG/DSAPPD) Data science (big data) application design, including languages for big data (Python, R), tools and models for data presentation and visualisation

- KA02.07 (DSENG/IS) Information systems, to support data-driven decision-making, with a focus on data warehouse and data centres

The DS-BoK provides mapping of the DS-BoK to existing classifications and BoKs: ACM Computer Science BoK (CS-BoK) selected KAs [25], Software Engineering BoK (SWEBOK) [28], DAMA Data Management BoK (DMBOK) [30] and related scientific subjects from CCS2012 [14]: computer systems organisation, information systems, software and its engineering.

6.2.2.1   DSENG/BDIT: Big Data Infrastructure Technologies Course Content

Big data infrastructures and technologies shape many of the data science applications. Systems and platforms behind big data differ significantly from traditional ones due to specific challenges of volume, velocity and variety of data that need to be supported by data storage and transformation. Data Lakes and SQL/NoSQL databases must be included in the DSENG curriculum.

Deployment of data science applications is usually tied to one of the most common platforms, such as Hadoop or Spark, hosted either on private or public clouds. The applications workflow must be linked to a whole data processing pipeline including ingestion and storage for a variety of data types and sources. Data scientists should have a general understanding of data and application security aspects in order to properly plan and execute data-driven processing in the organisation. This module should provide an overview of the most important security aspects, including accountability, compliance and certification.

6.2.2.2   Data Management and Data Stewardship in the Big Data and Data Science Curriculum

Data management and governance (DMG) [30, 77], although belonging to different KAG4-DSDM, must accompany the DSENG courses, and a short overview of the DMG common practices must be included into the BDIT curriculum. This should also include the introduction of the FAIR data principles (data must be findable, accessible, interoperable, reusable) [78] that are growingly adopted by the research community and recognised by industry. Data stewardship is a DMG application domain that combines general and subject domain data management ensuring that the FAIR principles are incorporated into the organisational practice.

6.2.3   Platforms for Big Data Processing and Analytics

This section describes what platforms can be used for teaching the BDIT4DA course and other courses in the data science engineering curricula requiring processing of big data. The section describes the Hadoop Ecosystem and its main components and functionalities and provides information about cloud-based Big data infrastructure and analytics platforms from the major cloud providers.

6.2.3.1   Essential Hadoop Ecosystem Components

Hadoop is commonly used as the main platform for big data processing; it includes multiple components and applications developed by the Apache Open Source Software community, with rich functionality to support all processes and stages in the data processing workflow/pipeline. Giving general understanding and basic experience with the Hadoop applications and tools is an important part of the practical activity and assignments in the BDIT4DA course. Figure 6.5 illustrates the Hadoop main components and few other popular applications for data processing [79, 80].

The following main Hadoop applications constitute the foundation of the Hadoop ecosystem and provide a basis for other applications:

- *HDFS*: Hadoop Distributed File System optimised for large-scale storage and processing of data on commodity hardware
- *MapReduce*: A YARN-based system for parallel processing of large datasets
- *YARN*: A framework for job scheduling and cluster resource management
- *Tez*: A generalised data-flow programming framework, built on Hadoop YARN, which provides a powerful and flexible engine to execute an arbitrary DAG of tasks to process data for both batch and interactive use cases.

Other Hadoop-related projects at Apache that provide a rich set of functionalities for data processing during the whole data lifecycle:

- *Hive*: A data warehouse system that provides data aggregation and querying

Fig. 6.5 Main components of the Hadoop ecosystem

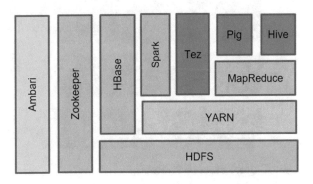

- *Pig*: A high-level data-flow language and execution framework for parallel computation
- *HBase*: A distributed column-oriented database that supports structured data storage for large tables
- *Spark*: A fast and general compute engine for Hadoop data. Spark provides a simple and expressive programming model that supports a wide range of applications, including ETL, machine learning, stream processing and graph computation
- *Mahout*: A scalable machine learning and data mining library
- *Solr*: Open source enterprise search platform that uses Lucene as indexing and search engine
- *Oozie*: Server-based workflow scheduling system to manage Hadoop jobs
- *Ambari*: A web-based tool for provisioning, managing and monitoring YARN jobs and Apache Hadoop clusters
- *Hue*: A user graphical interface providing full functionality for programming Hadoop applications, including dashboard, data upload/download, visualisation.

6.2.3.2 Hadoop Programming Languages

Introducing multiple Hadoop programming options is essential to allow future integration of the Hadoop platform and tools into research and business applications. Hadoop is natively programmed in Java, with current support for Scala by many applications. There is also support for Hadoop API calls from many popular programming and data analytics IDE and tools for R, Python, C, .NET. Specific for Hadoop are query languages to work with HBase, Hive, Pig as shown in Fig. 6.6.

Hive Query Language (HiveQL or HQL) [81]: Provides higher level data processing language, used for data warehousing applications in Hadoop. Query

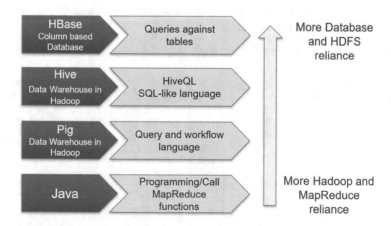

Fig. 6.6 Query languages for Hadoop

language is HiveQL, variant of SQL; tables are stored on HDFS as flat files. HiveQL facilitates large data processing that compiles down to Hadoop jobs.

Pig Latin [82] is a scripting language used for large-scale data processing system to describe a data processing flow. In fact, Pig Latin has similarity to HiveQL query commands with additional flow control commands. Similar to HiveQL, it compiles down to Hadoop jobs and relies on MapReduce or Tez for execution.

6.2.3.3   Cloud-Based Big Data Platforms

Major cloud platforms Amazon Web Services (AWS) [83], Microsoft Azure [84] and Google Cloud Platform (GCP) [85] provide a rich set of the big data services and applications:

- *AWS big data stack*: includes such services as Elastic MapReduce (EMR) which is a hosted Hadoop platform for data analytics; Amazon Kinesis is a managed service for real-time processing of streaming big data; Amazon DynamoDB— highly scalable NoSQL data stores; Amazon Aurora—scalable relational database; and Amazon Redshift—fully managed petabyte-scale data warehouse. Separately provided is the machine learning stack with a number of services. All services and tools are accessible from the AWS Console and can programmed via command line interface (CLI), where the former provides all necessary functionality to program, deploy and operate complex business applications by integrating all necessary components into one data processing pipeline.
- *Microsoft Azure*: provides well-integrated and supported by development tools the big data and analytics stack that includes such services as HDInsight which is Hortonworks-based Hadoop platform, Data Lake Storage and Data Lake Analytics, CosmosDB mufti-format NoSQL database and other services.
- *Google Cloud*: provides general cloud services and a set of easy configured big data services such as BigQuery column-based NoSQL database, Google Spanner Big SQL database and machine learning stack with well-defined APIs that support the whole data analytics.

6.2.4   Example Big Data Infrastructure and Technologies for Data Analytics Courses and Experience

This section provides an example of three Big Data Infrastructure and Technologies for Data Analytics, BDIT4DA, courses that can be adjusted to different academic or training programmes. BDIT4DA includes lectures, practice/hands-on labs, projects and such engaging activities as literature study and seminars. The courses should beneficially include a few guest lectures, to expose the students to external experts and real practices.

6.2.4.1 BDIT4DA Course for Big Data Engineering Master's

6.2.4.1.1 BDIT4DA Lectures

Lectures must provide a foundation for understanding the whole BDIT4DA technology domain, available platforms and tools and link other course activities. However, form and technical level must be adjusted to the incumbent programme, for example distinguishing computer science and MBA programmes. The same should be related to the selection of practical assignments and used tools and programming environment.

The following example is the set of lectures that has been developed and taught by the authors. Depending on program configuration and scheduling mentioned below, topics can be delivered in the form of sessions that can combine lectures (2–3 h), practice (2–4 h) and interactive activities such as literature review and project progress presentation. The following are the lectures:

- Lecture 1: Cloud computing foundation and economics: Cloud service models, cloud resources, cloud services operation, multitenancy. Virtual cloud data centre and outsourcing enterprise IT infrastructure to cloud. Cloud use cases and scenarios for enterprise. Cloud economics and pricing model.
- Lecture 2: Big data architecture framework, cloud-based big data services: big data architecture and services. Overview of major cloud-based big data platforms: AWS, Microsoft Azure, Google Cloud Platform (GCP). MapReduce scalable computation model. Overview of Hadoop ecosystem and components.
- Lecture 3: Hadoop platform for big data analytics: Hadoop ecosystem components: HDFS, HBase, MapReduce, YARN, Pig, Hive, Kafka, others.
- Lecture 4: SQL and NoSQL databases: SQL basics and popular RDBMS. Overview of NoSQL database types. Column-based databases and their use (e.g. HBase). Modern large-scale databases AWS Aurora, Azure CosmosDB, Google Spanner.
- Lecture 5: Data streams and streaming analytics: Data streams and stream analytics. Spark architecture and components. Popular Spark platforms, DataBricks. Spark programming and tools, SparkML library for machine learning.
- Lecture 6: Data management and governance and stewardship: enterprise big data architecture and large-scale data management. Data governance and data management. FAIR principles in data management.
- Lecture 7: Big data security and compliance: big data security challenges, data protection, cloud security models. Cloud compliance standards and cloud provider services assessment. CSA Consensus Assessment Initiative Questionnaire (CAIQ) and PCI DSS cloud security compliance.

Figure 6.7 illustrates in the form of 2D map relations between the proposed lecture and practice topics and the big data systems and applications infrastructure components. Using this kind of illustration provides a good guidance for designing

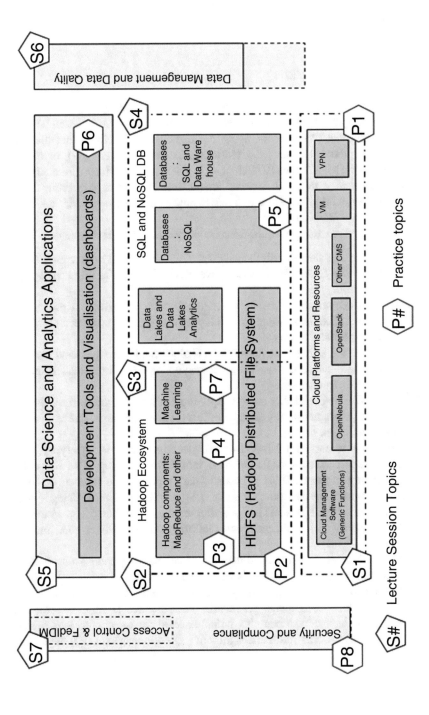

Fig. 6.7 BDIT4DA course lectures and practice topics map

courses with better practical orientation and at the same time providing advice to students for future self-study.

6.2.4.1.2   BDIT4DA Practice

Recommended practice includes working with the main Hadoop applications and programming simple data processing tasks. Different Hadoop platforms can be used for running practical assignments using dedicated Hadoop cluster installations (e.g. Cloudera Hadoop Cluster [86], Hortonworks Data Platform [87], or cloud-based AWS Elastic MapReduce (EMR), or Azure HDInsight platform). Students can be also recommended to instal personal single host Hadoop cluster using either Cloudera Starter edition or Hortonworks Sandbox that are available for both VirtualBox and for VMware.

The following are example topics for practice and hands-on assignments.

- Practice 1: Getting started with the selected cloud platform. For example, Amazon Web Services cloud; cloud services overview EC2, S3, VM instance deployment and access.
- Practice 2: Understanding MapReduce, Pregel, other massive data processing algorithms. Word count example using MapReduce algorithm (run manually and with Java MapReduce library).
- Practice 3: Getting started with the selected Hadoop platform. Command line and visual graphical interface (e.g. Hue), uploading, downloading data. Running simple Java MapReduce tasks.
- Practice 4: Working with Pig: using simple Pig Latin scripts and tasks. Develop Pig script for programming Big Data workflows. This can be also done as a part of practical assignment on Pig.
- Practice 5: Working with Hive: Run simple Hive script for querying Hive database. Import external SQL database into Hive. Develop Hive script for processing large datasets. This can be also a part of practical assignment on Hive.
- Practice 6: Streaming data processing with Spark, Kafka, Storm. Using simple task to program Spark jobs and using Kafka message processing. The option for this practice can also use Databricks platforms that provide a good tutorial website.
- Practice 7: Creating dashboard and data visualisation. Using tools available from the selected Hadoop platform to visualise data, in particular using results from Practice 5 or 6 that is dealing with large datasets where dashboard is necessary.
- Practice 8: Cloud compliance practicum. This practice is important for the students to understand the complex compliance issues for applications run on cloud. Using Consensus Assessment Initiative Questionnaire (CAIQ) tools.

6.2.4.2   Course for Data Science Masters

In contrast to the big data engineer example, a course for data scientists spends more time on algorithm design aspect. All basic tool and concepts are introduced, but less time is spent on topics related to security and governance.

First five lectures have corresponding laboratory sessions. Afterwards, students begin on working on group projects on datasets of their choice, applying concepts, technologies and tools from lectures. Progress in projects is presented at plenary presentations sessions, in the middle and at the end of the course.

To further motivate and guide students, 1 or 2 guest lectures with practitioners from industry are organised. They can be scheduled any time after Lecture 5. In some case, it might also be scheduled together with Lecture 1.

6.2.4.2.1   Lectures

- Lecture 1: Introduction to data-intensive systems and use cases. Data as fourth paradigm of science. Increasing focus on data collection, data architectures, data centrs. Use cases in search, commerce, healthcare, energy.
- Lecture 2: Hadoop 101 and functional abstraction. Introductory, but fully functioning MapReduce program in Python with execution from command line.
- Lecture 3: MapReduce. Detailed description of file splitting, mapping, combining, shuffling, reducing and storage of results.
- Lecture 4: Hadoop architecture. Resource management, permanent and temporary storage, batch processing, real-time processing, higher-level tools.
- Lecture 5: MapReduce algorithms and patterns. Counting, summing and averaging. Processing multiline input. Random sampling. Search Assist. Inverted index.
- Lecture 6: HBase and other NoSQL databases. Alternative permanent storage for big data. CAP/PACELC theorems. Interaction between Hadoop/MapReduce and NOSQL databases.
- Lecture 7: First project presentation. Focused on choice of dataset, data preprocessing, identification of interesting problems.
- Lecture 8: Spark (RDD based). Data model. Programming model, actions, transformations, other operations. Architecture.
- Lecture 9: Spark (SQL/other structures/MLlib). Alternative programming models, advantages and drawbacks. Incorporating existing libraries in the programming workflow.
- Bonus: 1–2 industrial guest lectures. Usually focusing on data quality and data workflow in industry.
- Lecture 10: Final project presentations. Focused on MapReduce implementation of identified problems. Performance tuning.

6.2.4.2.2   Practice and Project Development

- Lab 1: Refresh Bash knowledge, set up Docker and Hortonworks Sandbox. Ensures that students have a working test environment on their laptops.
- Lab 2: Recreate steps from lecture (system setup, file coping, running ready MRJob and Hive examples). Ensures that students can correctly execute examples in the book/lecture.
- Lab 3: Introduce modifications to MRJob-based program on the Sandbox. Ensures that students understood basic concepts related to MapReduce programming.
- Lab 4: Set up Hadoop from scratch on a VM (not Sandbox).
- Ensure that students understood Hadoop architecture.
- Lab 5: In-depth analysis of typical algorithms and patterns in groups. Ensures that students understood details of MapReduce programming.

After five laboratory sessions students work on group projects. They are still encouraged to come on a regular basis to laboratory sessions where they can discuss and get support with any technical problems they face.

6.2.4.3   Big Data Infrastructure Technologies (BDIT) Course for MBA in Big Data

Big data and data analytics tools is important part of the business supporting infrastructure and services which are growingly cloud based. The specifics of the MBA data science groups is the diverse background of the students from the economics and business to computers science and engineering. The main goal of the BDIT course is to provide knowledge sufficient for the future business managers to make assessment and advise development of necessary services in their future organisations. The practical work is entirely based on using cloud-based applications and tools. The course includes also projects where the students working in groups need to deliver the design of the cloud-based big data infrastructure supporting the business processes of their hypothetical company.

6.2.4.3.1   Lectures

BDIT lectures include subset of topics outlined in Sect. 6.2.1 but enriched with examples and closely linked to practices and labs.

- Lecture 1: Cloud computing foundation and cloud economics: Provides basics for understanding and working with clouds.
- Lecture 2: Big data architecture framework, cloud-based big data services: Overview of cloud-based big data platforms and tools, including AWS, Azure and Google Cloud Platform.

- Lecture 3: MapReduce and Hadoop platform: Introduce the Hadoop ecosystem and main components; example of use Lecture 4 Spark and Streaming Analytics: Including data structure, programming with Scala.
- Lecture 5: SQL and NoSQL databases: Database classification and types, cloud-based big databases, Hadoop-based HBase, Hive
- Lecture 6: Data management and governance: based on DAMA DMBOK, extended with FAIR and QA methods
- Lecture 7: Big data security and compliance: Cloud data security services, access control, CSA Compliance framework.

6.2.4.3.2   Practice and Project Development

Practice covers major aspects of working with two main cloud platforms AWS and Microsoft Azure, starting with AWS as presenting more generic cloud services model, and following with Microsoft Azure as providing better aligning with business analytics processes. The following topics were included in the course:

- Practice 1: Getting started with Amazon Web Services cloud
- Practice 2: AWS services EC2, S3 deployment and access
- Practice 3: Amazon Elastic MapReduce (EMR). Running MapReduce word count example manually and using EMR
- Practice 4: AWS Aurora scalable SQL database, deployment and simple queries exercises
- Practice 4: Getting started with Microsoft Azure cloud, storage and compute services, instances deployment
- Practice 5: Azure HDInsight business analytics platform, deployment and Hadoop cluster visual interface. Running simple Spark examples
- Practice 6: Cloud compliance practicum using CSA Consensus Assessment Initiative Questionnaire (CAIQ) tools

6.2.5   Conclusions

The above-described BDIT4DA course has been taught by the authors in different programs and different installations: campus face-to-face teaching, part-time evening lecturing and practice and remote lecturing. Experience confirms that in general lecture materials can be used the same, given that there is no single textbook for the course. However, practice must be adopted to the hosting master's programme, student background and lecture-practice scheduling. Important aspect of this course and any other course related to data science is to develop in students a kind of data-centric approach and thinking. This aspect is related to the data science professional skills ("Thinking and acting like data scientist") which are defined in the EDISON Data Science Framework and discussed in the authors' publications [64, 65, 76].

The presented general approach and practical experience in teaching the Big Data Infrastructure Technologies for Data Analytics is based on the EDISON Data Science Framework, which is widely used by universities, professional training organisations and certification organisations, providing valuable feedback for further framework development and continuous courses evolution. The presented work is also based on the long author's experience in teaching cloud computing technologies [88] that provide computational and infrastructure basis for the big data technologies. The cloud computing curriculum design used the technology maps and linked professional profiles, together with the Bloom's taxonomy, to design the customisable curriculum. Such approach has been developed into the EDSF and its main components.

The academic education or professional training must provide a strong basis for graduates and trainees to continue their further self-study and professional development in conditions of the fast-developing technologies and agile business environment adopted by the majority of modern companies. To achieve this, the data science curriculum needs to be supported by the professional skills development courses such as to develop the general twenty-first-century skills and specific data science workplace skills.

One of general skills for data workers is considered the data management and governance and specifically the research data management and stewardship adopting FAIR data principles, which is a part of the authors' cooperation in the FAIRsFAIR project [58].

The EDSF maintenance and continuous development as well as collection of the best practices in data science education and training is supported and coordinated by the EDISON community, in cooperation with national and EU projects as well as supported by the Research Data Alliance (RDA) Interest Group on Education and Training on Handling Research Data (IG-ETHRD) [5]. Participation and contribution to both the IG-ETHRD and EDSF Community Initiative is open and free.

6.3   Big Data Value Data Science Badges

Ernestina Menasalvas, Ana M. Moreno, and Nik Swoboda

6.3.1   Introduction

With the development of new technology and the digital transformation of our economy, the labour market has also evolved. Nowadays, applicants for a job are no longer asked to submit a traditional paper resume; this information is presented digitally; recruiters and headhunters search the Internet (on an international level) for candidates who have the required skills for their needs; and some assessment of candidates can be done online. Moreover, the demands for labour are constantly evolving and the required skills and qualifications change rapidly over time. Adequately adapting to these changes is essential for the success of employers, learning institutions and governmental agencies related to education.

In this chapter, we will first discuss mechanisms for recognising skills in the EU with a focus on the internationalisation, digitalisation and flexibility of those credentials. Then we will consider their application to data science with the goal of proposing a new framework for the recognition skills in data science. We begin with a brief review of the main challenges we hope to address.

How can we standardise credentials throughout Europe?

Although political institutions in the EU have strived to coordinate and standardise diplomas and other forms of credentialing in higher education, the variety of educational systems in the EU and the lack of an adequate system to recognise learning and skills have contributed to great differences in the economic and social outcomes of the member states. The many different educational and training systems in Europe make it difficult for employers to evaluate the capabilities of potential employees.

Currently, there is no automatic system for the EU-wide recognition of academic diplomas [89]; students can only request a "statement of comparability" for their university degree. This statement of comparability details how the student's diploma compares to the diplomas of another EU country [90]. Something similar happens with the recognition of professional qualifications: the mobility of Europeans between member states of the EU often requires the full recognition of their professional qualifications (training and professional experience). This is accomplished through an established procedure in each European country [91]. Directives 2005/36/EC and 2013/55/UE on the recognition of professional qualifications establish guidelines that allow professionals to work in another EU country, different from the one where they obtained their professional qualification, on the basis of a declaration.

These directives provide three systems of recognition:

- Automatic recognition—for professions with harmonised minimum training conditions, i.e. nurses, midwives, doctors, dentists, pharmacists, architects and veterinary surgeons
- General system—for other regulated professions such as teachers, translators and real estate agents
- Recognition on the basis of professional experience—for certain professional activities such as carpenters, upholsterers and beauticians

Additionally, since 18 January 2016 the European professional card (EPC) has been available for five professions (general care nurses, physiotherapists, pharmacists, real estate agents and mountain guides). It is an electronic certificate which can be applied for online.

Unfortunately, these existing mechanisms do not easily accommodate many professions including that of data science.

How can data science credentials be: digital, verifiable, granular, and quickly evolving?

Traditionally skills and credentials were conveyed via a resume on paper and other paper-based credentials. Nowadays, this information can be shared via the Internet in web pages, in social media and in many other forms. The digitalisation of credentials not only allows easier access but also offers new possibilities like:

- The online verification of the validity of the credentials
- Greater granularity in the definition of the credentials
- The expiration of credentials requiring their periodic renewal which could take into account changes in the demands for skills
- Access to the evidence used in the awarding of credentials

Future schemes for the recognition of skills need to adapt to and accommodate these new demands.

Overview

This section begins with a summary of current trends regarding education and skills in Europe. From these trends, we extract a series of desirable properties for a data science skills recognition scheme. Then, a survey of different mechanisms for the recognition of skills is presented. Next, a comparison and critical analysis of these recognitions is provided while taking into account the previously identified desirable properties. Lastly, a recommendation for a data science skills recognition process is proposed.

6.3.2 The Strategic Framework for Education and Training

Even though the educational systems of each country in the EU are managed separately at the national level, a common EU policy exists to support those systems and to help address common challenges faced by the EU. The Strategic Framework for Education & Training 2020 (ET-2020) contains the current policies of the EC for cooperation on education and training. These policies were initially adopted in 2009 [92] and contain four common objectives that should be met by 2020 in the EU:

- Making lifelong learning and mobility a reality
- Improving the quality and efficiency of education and training
- Promoting equity, social cohesion and active citizenship
- Enhancing creativity and innovation, including entrepreneurship, at all levels of education and training

In reviewing the progress of the ET-2020, the *2015 Joint Report of the Council and the Commission on the implementation of the strategic framework for European cooperation in education and training* [93] (2015-JR-SFECT) included a new set of "priority areas for European cooperation in education and training":

1. Relevant and high-quality skills and competences for employability, innovation, active citizenship
2. Inclusive education, equality, non-discrimination, civic competences
3. Open and innovative education and training, including by fully embracing the digital era
4. Strong support for educators
5. Transparency and recognition of skills and qualifications
6. Sustainable investment, performance and efficiency of education and training systems

As the emphasis on promoting the "transparency and recognition of skills and qualifications" is particularly relevant to the task of recognising data science skills, we will focus further on that priority area. To explain this priority area, the report identifies these concrete needs:

- Fostering transparency, quality assurance, validation and recognition of skills and/or qualifications, including those acquired through digital, online and open learning and the validation of informal and non-formal learning
- Simplifying and rationalising the transparency, documentation, validation and recognition tools that involve direct outreach to learners, workers and employers and further implementing the EQF (European Qualifications Framework for Lifelong Learning) and NQFs (National Qualifications Frameworks)
- Supporting the mobility of pupils, apprentices, students, teachers, members of educational staff and researchers
- Developing strategic partnerships and joint courses, in particular through increasing internationalisation of higher education and vocational education and training

To help reach these goals the EU has promoted several programmes:

- *The European Qualifications Framework for Lifelong Learning (EQF)* supports the process of validating qualifications by providing a common reference for qualification levels throughout Europe and the linking of member state validation systems with formal qualifications systems.
- *Validation of non-formal and informal learning* is the process of recognising an individual's knowledge, skills and competences gained outside formal educational systems. To help achieve this recognition, the CEDEFOP (European Centre for the Development of Vocational Training) in cooperation with the EC has defined some guidelines for validating non-formal and informal learning [94]. They have also proposed an up-to-date European inventory that provides an overview for each country and good practices for the design and implementation of validation initiatives.
- The *Europass portfolio* also relates to validation systems because it documents learning and enables users to display their skills, qualifications and experiences in a uniform way across Europe.
- The *European Credit Transfer and Accumulation* System for higher education [95] and the *European Credit system for Vocational Education and Training*

(ECVET) [96] standardise the quantification of formal learning throughout Europe, making it easier to compare the time spent in educational programmes.

• *Quality assurance* arrangements in higher education and vocational training to make education systems easier to understand for students and employers, by improving transparency tools.

We will now elaborate more upon the most relevant points of a number of key initiatives within this framework.

6.3.2.1   New Skills Agenda for Europe

The New Skills Agenda for Europe (NSAE) is one of the initiatives of the EU to help meet the targets of the ET-2020. On 10 June 2016, the European Commission presented this agenda [97], which sets out different guidelines to guarantee that the most adequate training, skills and career guidance is accessible to everyone in the EU.

The New Skills Agenda emphasises "the strategic importance of skills for sustaining jobs, growth and competitiveness" and is focused on three key points:

• Improving the quality and relevance of skills formation
• Making skills and qualifications more visible and comparable
• Improving skills intelligence and information for better career choices

The most relevant concerns and recommendations mentioned in the New Skills Agenda regarding data science skills in Europe include:

• Future credentials should easily allow the comparison of students' skills through-out the EU.
• Both the employed and the unemployed need adequate ways to present their skills and qualifications. Employers need ways to identify and recruit new employees with the skills that they need.
• Once skills qualifications are easily accessible, current and future demands for skills could be identified with data science analysis (skills intelligence).
• Current qualification systems focus on the learning outcomes of formal education programmes, but do not validate non-formal and informal learning. Ongoing learning, including learning at the workplace, needs to be encouraged.
• Skills acquisition should not only be in formal education and training (literacy, numeracy, science, foreign languages) but also transversal skills (teamwork, creative analysis, problem-solving, entrepreneurship, etc.).

6.3.2.2   New Europass Framework

In February 2005, the Europass was launched, following the decision of the European Parliament and the Council [98] to create a single framework whose goal was to make individuals' skills and qualifications more comprehensible in

Europe. This was done in the interest of facilitating the mobility of students and workers. The Europass consisted of a portfolio of five documents: the Europass curriculum vitae (CV), the Europass language passport, the Europass mobility, the Europass certificate supplement and the Europass diploma supplement.

Despite the fact that the European CV has undergone significant improvements to adapt to the changes brought by the technological revolution, the initial Europass did not address the changing educational, training and labour market conditions [99]:

- It focused on documents and templates that are not compatible with the use of social media, mobile devices, big data analysis and job matching tools.
- It did not face the growing relevance of modern learning, which needed an easy way to record skills and qualifications acquired through non-formal or informal learning, including online learning.
- It did not take into account the use of tools such as "open badges".

On 4 October 2016, the Commission decided to revise the Europass Decision, by building a new Europass framework that contributes to the display of people's skills and qualifications in a unified manner for all EU countries [100].

The proposal addresses challenges regarding the way that information technology has changed the labour market and new educational possibilities:

- The publication of employment offers, job applications, candidate's evaluation and recruiting are increasingly done online through tools that use social media, big data and other technologies, making it easier to find information on skills and qualifications.
- Education and training is increasingly offered online using digital platforms; at the same time, skills, experiences and learning achievements (formal and non-formal) are recognised in different forms, such as open badges.

With the new framework users can display their skills and qualifications in new formats, as the revised Europass uses open standards to facilitate the exchange of electronic data and defines authentication measures to ensure the validity of the digital content.

To achieve this goal, the new infrastructure includes several new tools, which allow users to give evidence of their skills and qualifications in all EU languages; these tools are:

- An online tool to create persona profiles, including both the traditional CV, with work experience and training/education and the skills recognition
- Applications to help evaluate the users' skills
- Information on learning opportunities across Europe
- Assistance on how to get a user's skills recognised
- Labour market intelligence, to learn which skills are more valuable

The new Europass is connected to other EU tools and services related to work, education and training systems, to encourage the exchange of information and to help users in their education and career path decisions.

6.3.2.3   European Qualifications Framework for Lifelong Learning

In April 2008, the European Parliament and Council resolved to establish the European Qualifications Framework for Lifelong Learning [101]. The process was voluntary, and countries were invited to implement the framework in two stages: the first, which was to be completed by 2010, relating national qualification levels to the EQF; and the second, by 2012, ensuring that all new qualifications issued in Europe include references to the appropriate EQF level.

The aim of this framework is to provide a way to compare and interpret the levels of different qualifications in the EU and thereby make those qualifications more transparent. This also facilitates mobility, having positive effects for learners and workers, who can have their level of competence recognised using a standard description all across Europe. This proposal is also beneficial for recruiters and education providers, who will be able to understand the applicants' qualifications. The adoption of a common reference framework eases the comparison and recognition of traditional qualifications issued by national authorities and those awarded by third parties (e.g. multinational companies). This allows the comparison of formal and non-formal education by increasing the transparency of qualifications awarded outside the formal education system.

The EQF can be applied to any kind of education, training or qualification including required basic education to advanced academic and professional and training. It consists of eight qualification levels, given in terms of learning outcomes: knowledge, skills and competences. These levels take into consideration theoretical knowledge, practical skills, technical skills and social competences.

However, the adoption process of the EQF has proved difficult: differences have appeared when comparing general education certificates in different national systems with the EQF levels. For example, for a similar school certificate, some countries assign a level 2 or 3 (for secondary education) and others a level 4 or 5 (for higher education). This same problem has occurred with vocational education.

6.3.2.4   European Skills, Competences, Qualifications and Occupations

The European Skills, Competences, Qualifications and Occupations (ESCO) framework is a multilingual classification system that aims to bridge the communications gap between industry and those offering training through a common reference terminology. The ESCO initiative was launched in 2010 by the EU and the first version of the ESCO framework was published in July 2017.

The ESCO benefits those in the labour market and in the education and training sector in a variety of ways:

• By providing a better matching of people to jobs by employment services or electronic tools:

- Helping employers define the set of skills, competences and qualifications for a vacant job
- Helping job seekers build professional profiles in a terminology that suits job vacancies
- Enable mobility through Europe

- By supporting education and training systems in the move to learning outcomes that better meet labour market needs:

  - Supporting the provision of information to education and training institutions that can help them in the development of new curricula
  - Helping to provide more transparent information to students on learning outcomes and the relevance of qualifications to the labour market before they commence education or training

By supporting evidence-based policymaking:

- Enhancing the "collection, comparison and dissemination of data in skills intelligence and statistics tools, among others, in the European Skills Panorama" [102].

The ESCO is built upon three pillars: occupations, skills/competences and qualifications.

Regarding qualifications, the ESCO is based on the EQF framework and the national databases that most member states developed or are developing, in which they assign an EQF level to each qualification and describe the expected outcome.

The most relevant ESCO guiding principles are:

- Useful, ESCO aims to become the de facto standard of the identification of occupations, skills competence and qualifications.
- Accepted, ESCO aims to be voluntarily adopted by stakeholders.
- Updated, ESCO will be continuously updated and adapted.
- Flexible, ESCO does not aim to standardise the scope of occupations but to provide standard terminology.
- High-quality, different stakeholders carefully ensured the quality of the ESCO.
- Transparent and open development, results were shared with interested parties and it was open to all stakeholders.
- Machine readable and compatible with existing IT systems and standards.

6.3.2.5   Highlights and Common Threads in These Initiatives

After reviewing these political trends in Europe, we can now extract some of the key properties that a data science skills recognition programme should contain in order to be in agreement with these existing efforts.

A data science skills recognition system should:

- P1 (2015-JR-SFECT, NSAE, Europass, EQF, ESCO) be transparent, accessible and allow the easy comparison of students' skills throughout the EU
- P2 (2015-JR-SFECT, ESCO) include an assurance of quality
- P3 (2015-JR-SFECT) provide tools for their verification and validation
- P4 (2015-JR-SFECT, NSAE, Europass, EQF) include skills acquired through traditional, digital, online and open learning, as well as the validation of informal and non-formal learning
- P5 (2015-JR-SFECT, EQF, ESCO) be compatible with the EQF
- P6 (NSAE) influence the relevance of the skills being acquired
- P7 (NSAE, Europass, ESCO) allow the digital analysis of both the demand for and the availability of skills
- P8 (Europass) allow their use online: in platforms like the Europass, social media and on mobile devices
- P9 (EQF, ESCO) focus on learning outcomes and not on traditional measures such as hours of study

With these goals in mind, we will now look at the most common and popular methods of recognising skills.

6.3.3 Education and Training Recognitions

6.3.3.1 A Survey of Recognitions

Accreditations

"Accreditation [is] the formal recognition by an independent body, generally known as an accreditation body, that a certification body operates according to international standards." [103]

In the context of higher education in Europe, accreditation is the process by which an educational programme acquires the right to grant degrees. In Europe, most accreditation agencies are endorsed by national governments and accredit all of that country's degrees. Sometimes these agencies recognise European or international accreditations and simplify the national accreditation process for programmes already accredited at the European or international Level.

Requirements for accreditation vary depending on the accreditation agency. Some examples of accreditation agencies include the Agencia Nacional de Evaluación de la Calidad y Acreditación in Spain, the UK Accreditation Service in the UK, the Accreditation Board for Engineering and Technology or the World Association of Conformity Assessment Accreditation Bodies.

University/Academic Degrees

Of the collection of recognitions we will review, this is most certainly the one with the longest history. For example, the notion of a doctorate was established in medieval Europe and was considered to be a licence to teach at the university level [104]. Nowadays, the European system of university degrees (bachelor's degree, master's degree and the doctorate) are used worldwide.

Accredited college and university programmes have the right to award academic degrees. It should be noted that in some parts of the world, unaccredited programmes (sometimes referred to as degree mills) can legally confer degrees, but these degrees are often considered to be of little worth [105].

A great variety of requirements exist but typically 4 years of university study must be completed to be awarded a bachelor's degree, two additional years of study for a master's degree and a significant research contribution is required for the awarding of a doctorate.

In many private and public sector jobs, both pay scales and position prerequisites are directly related to the degrees held by a candidate. In some parts of the world, holding a degree results in a change of title (Doctor for example) and in others it results in the right to use post-nominal letters (BA for example).

Noteworthy examples in big data/data Science include M.Sc. Big Data & Business Analytics, University of Amsterdam; M.Sc. Applied Informatics, Vytautas Magnus University; M.Sc. Data Science, Sapienza Universita di Roma; or M.Sc. Computer Science, National University of Ireland.

Certificates

A certificate is simply a document that attests to the fact that a certain individual has "received specific education or has passed a test or series of tests [106]". Though in reality they are very varied in use, in the context of the computing industry certificates are most commonly used to recognise knowledge regarding a specific set of skills.

There are both academic and professional certificates with the former being awarded by higher education providers while the latter are awarded by professional organisations or individual companies related to their own products.

Certificates typically require less effort to obtain than an academic degree. The examples given below have a duration of between 6 and 18 months. Many certificate programmes are specifically designed for "continuing education" students who are already employed full-time but are trying to advance their career. Many academic certificates are associated with traditional coursework and are basically equivalent to having passed with a certain grade a set of courses. Most professional certificates require passing a test; some also require a certain amount of professional experience to be eligible for certification. Many professional certificates expire after a certain period of time and require renewal by completing continuing education courses and/or exams.

Corporations often require that service providers have staff with certain certifications before they can provide specific services. Often job advertisements specifically require certain certifications. For example, many network engineer positions require that applicants have a Cisco Certified Network Associate (CCNA) certificate.

Relevant examples of academic certificates in big data/data science include:

- Harvard Data Science Certificate [107]: Cost: 11,360 USD, duration: 1.5 years; all courses can be completed online

- University of California, Irvine Data Science Certificate Program [108]: Cost: free for UCI graduate students, duration: 32 h of courses/workshops
- Georgetown University Certificate in Data Science [109]: Cost: 7496.00 USD, duration: 6 months, courses held on Friday evenings and Saturdays
- UC Berkeley Certificate Program in Data Science [110]: Cost: 5100 USD (+ materials + registration), duration: 150 h of class

Labels

Labels are a distinction awarded to an existing degree programme. The use of the term label appears to be almost exclusively European. Like accreditations, but unlike the other recognitions mentioned here, it is applied to the programme itself and is not normally thought of as being applied to the programme's participants (except through association). A label publicly recognises that the programme in question meets the requirement of the label issuer.

There is no clear consensus regarding who can confer a label, but at the moment the most noteworthy labels are conferred by programmes funded by the European Commission. Each organisation is free to design any requirements, which it sees fit.

Noteworthy examples are Erasmus Mundus Master of Excellence, Erasmus+, EIT Digital, Eur-ace, Euro-Inf and the European Language Label (ELL).

Badges

Badges are a very recent arrival in the skills recognition landscape. A badge is a graphical representation of any kind of achievement, goal or milestone based on a digital file that integrates the criteria and evidence used to obtain the badge [111]. In an industrial setting, badges seem to be a lesser and more accessible form of recognition when compared with certificates. One of the principal motivations behind the "Badge Movement" was to provide a new mechanism for the recognition of skills which is better adapted to recent changes in learning:

- Nowadays, learning happens in many different contexts and while using various kinds of media. Instead of only recognising credentials from students enrolled in established learning institutions, employers should have a mechanism for recognising skills (and experiences) acquired by anyone through professional training programmes, participating in competitions, volunteer programmes, MOOC's, etc.
- Learning is not something which should only be "recognised" during the "student" phase of one's career but should rather be an ongoing process.
- Traditional skills recognitions do not capture elements which cannot be easily evaluated by test scores and short-term projects.
- Academic diplomas are for the most part monolithic documents. Recognitions of lesser granularity and much more flexibility are needed. Also, these recognitions should contain both the criteria used to evaluate the skills and evidence of the acquisition.
- Recognitions need to be digital, and capable of being displayed online.

By design, badges can be awarded by any organisation. Each organisation is free to design any set of requirements, which it sees fit.

Noteworthy examples are IBM Digital Badges [112], Digital Badges at Purdue University [113] or Stack overflow badge program [114].

The next section compares these recognition strategies according to the main characteristics identified from the EU political trends discussed in Sect. 6.3.2.

6.3.3.2   A Comparison of Recognitions

We began by reviewing current political trends in Europe regarding skills recognition and then we gave a brief overview of different popular skills recognition tools. The goal of this section is to evaluate the suitability of those recognition tools based upon those previously identified European political trends along with a short list of properties specific to data science. Lastly, based upon that evaluation we will recommend a scheme for the recognition of data science skills in Europe.

6.3.3.2.1   Properties Specific to Data Science

Before proceeding to the comparison, we would like to include a few additional properties not previously mentioned in Sect. 6.3.2.5 and which are specific to the rapidly evolving area of data science. Additionally, skills recognition in data science should:

- P10 require renewal after a set period of time
- P11 provide a framework which can quickly adapt to changes in skill requirement
- P12 measure skills on a highly granular and an individual-by-individual basis

6.3.3.2.2   The Comparison

It should be noted that many of the previously recognition tools are very flexible; thus, in certain cases they could or could not satisfy a certain property depending upon the implementation of the tool ("✓?")

Desired property	AC	UD	CE	LA	Badge
P1—transparent, accessible and allow the easy comparison of students' skills throughout the EU			✓	✓?	✓
P2—include an assurance of quality	✓	✓	✓	✓	✓?
P3—provide tools for their verification and validation	✓?	✓?	✓?	✓?	✓
P4—include skills acquired through traditional, digital, online and open learning, as well as the validation of informal and non-formal learning		✓?	✓	✓?	✓
P5—compatible with the EQF		✓?	✓	✓	✓
P6—influence the relevance of the skills being acquired		✓?	✓	✓	✓

(continued)

Desired property	AC	UD	CE	LA	Badge
P7—allow the digital analysis of both the demand for and the availability of skills			✓?		✓
P8—allow their use online: in platforms like the Europass, social media and on mobile devices		✓?	✓?	✓?	✓
P9—focus on learning outcomes and not on traditional measures such as hours of study		✓?	✓	✓	✓
P10—require renewal after a set period of time			✓		✓
P11—provide a framework which can quickly adapt to changes in skill requirement			✓	✓?	✓
P12—measure skills on a highly granular and an individual-by-individual basis			✓		✓

6.3.3.2.3 Discussion

In the previous comparison, the two tools which showed the worst results were accreditations and labels. Simply put, accreditation is a tool used to ensure that an educational programme meets some minimum requirements in order for it to issue degrees. As such, it is not suited for the recognition of data science skills. Labels have more promise but again are a tool used to recognise traditional educational programmes as a whole and not an individual's learning outcomes. University degrees have many strong points but fall short in that they:

- Are not very transparent nor individual
- Do not traditionally recognise informal and non-formal learning
- Are not very granular
- Are not traditionally digital
- Are not well suited to recognising quickly changing skills.

Both badges and certificates offer all of the desired properties and thus our recommendation will be a mix of both of these tools.

6.3.3.3 Recommendations for Data Science Skills Recognition

Given that both certificates and badges manifest the properties which we found desirable in a skills recognition tool, we propose a hybrid approach drawing from the strengths of both badges and certifications.

We begin with a summary of the needs of all stakeholders in the data science ecosystem.

Data scientists need:

- (DS-N1) Credentials, which are widely recognised
- (DS-N2) Credentials, which can be easily verified online

- (DS-N3) A simple way to digitally display their skills online and in social networks
- (DS-N4) Mechanisms to formally recognise skills acquired through informal and non-formal training

Employers who hire data scientists need:

- (EM-N1) Tools to verify the authenticity of credentials
- (EM-N2) A skills recognition framework, which facilitates the comparison of candidate skills throughout the EU
- (EM-N3) Influence in the process of designing the types of training data scientists receive
- (EM-N4) A scheme for recognising skills in data science, which can quickly adapt to changes in the data science ecosystem

Educators who train data scientists need:

- (ED-N1) Publicity for their programmes and the added value that an externally branded recognition of their training can provide
- (ED-N2) Recognitions for the partial completion of their programmes to assist students who are seeking employment while studying or students who abandon their studies
- (ED-N3) Contact with employers, a mechanism to clarify the changing needs of industry and clear recommendations regarding how to adapt to those needs

Accordingly, to meet these needs our recognition approach will be based on the following elements:

- OB. The use of open badges (DS-N2, DS-N3, EM-N1, EM-N4, ED-N2).
- E. Experts from industry and academia will contribute to both defining and maintaining the badge scheme (DS-N1, EM-N2, EM-N3, EM-N4, ED-N1, ED-N3).
     An initial proposal regarding both the types and the requirements of the badges will be based upon the work of the EDISON and EDSA projects. These projects have focused a great deal of time and effort in establishing a consensus regarding frameworks for data science education in both industry and academia.
- TP. Badges will only be issued by trusted third parties. By carefully vetting badge issuers and their practices the reputation of the credentials will increase (DS-N1, ED-N1).
- Pr. Supposing that the badges become popular, their requirements will influence the training of data scientists, give prestige to those who issue the badges and give publicity to their branding (EM-N3, ED-N1).
- InF. Include the recognition of skills acquired through informal and non-formal training (DS-N4).

Table 6.1 shows that our proposal covers all of the previously identified needs

Though what we are recommending is really a hybrid approach inspired by aspects of certificates and badges, given the novelty and interest by large

Table 6.1 Data science needs and recognition elements

	DS-N1	DS-N2	DS-N3	DS-N4	EM-N1	EM-N2	EM-N3	EM-N4	ED-N1	ED-N2	ED-N3
OB	•	•	•	•	•	•	•	•	•	•	•
E	•	•	•	•	•	•	•	•	•	•	•
Tp	•	•	•	•	•	•	•	•	•	•	•
Pr	•	•	•	•	•	•	•	•	•	•	•
InF	•	•	•	•	•	•	•	•	•	•	•

corporations such as IBM and Microsoft in the use badges, we believe that the term "badge" should be used for the data science skills recognition system.

The only disadvantage to the use of the term "badge" that we see is their current strong association with informal and non-formal learning. But in light of the growing interest, both in academia and in industry, to use badges we believe that in the medium term that association will decrease.

6.3.3.4 The Logistics of the BDV Data Science Badge Program

In this section, we will briefly describe the logistics of the BDV Data Science Badge Program.

Preliminaries:

- A collection of experts, including representatives from industry and academia, establish both the types and the requirements of the badges included in the programme. They also define the process for applying to issue badges.
- This group of experts also approves the members of a group of reviewers whose mission is to evaluate applications received to issue badges.

Applying to issue badges:

- Interested institutions/educators can apply to issue a badge. This application includes submitting evidence that shows that they provide their students with the skills required by the badge.
- Reviewers are assigned applications to assess and decide whether the applicant programme meets the established standards to issue badges. Applications can be rejected, accepted conditionally for 1 year or accepted for 4 years.

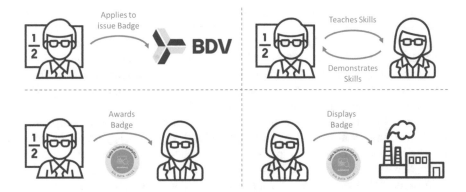

Fig. 6.8 BDV Badges—Application and Issuing Process

Issuing badges:

- Students in a data science programme authorised to issue BDV Badges acquire and demonstrate their data science skills through their studies. Students in the programme can submit an application to their programme to receive a badge.
- The programme reviews badge applications, and if the applicant has met the requirements of the badge, then it issues the student that badge. Badges are individualised and contain metadata including the requirements to earn the badge and evidence of the student's achievements.

Displaying and viewing badges:

- Students can display their badges online: in their CV, in social networks, etc.
- Interested employers can verify that a badge is valid and use its metadata to access relevant information regarding the earners of a badge.

Revision

- The group of experts will periodically meet to review the programme. Based upon progress reports, they can propose changes and improvements in the programme.

A graphical representation of the process for applying to issue and issuing badges is given in Fig. 6.8.

6.3.4   BDV Data Science Badges Types and Requirements

6.3.4.1   Initial Proposal

In order to define both the types and requirements of the BDV Data Science Badge Program we focused on EDISON's Data Science Competence Framework (CF-DS) and Data Science Model Curriculum (MC-DS). We selected these documents as the basis of our proposal as they directly relate competences with professional profiles. So from an industrial perspective, it should be easy to understand the value of the badges through their relationships to the corresponding competences. Furthermore, the professional profiles and their required competency levels as given by EDISON can be used as an example of how the badges can be used in each organisation. These examples can then be adapted to the structure and needs of an individual organisation to provide a mapping between the badges and its hiring needs.

We initially proposed the creation of one group of badges for each competence group, with each group of badges having three levels of proficiency (basic, intermediate and expert). To make the proposal more accessible to a wider audience, we chose to use the term "required skills" in place of "learning outcomes".

Thus, the following is the initial collection of BDV Data Science Badges:

- Data Science Analytics Badge
- Data Engineering Badge
- Data Science Management Badge
- Business Process Management Badge
- Data Science Research Method and Project Management Badge

We also considered the possibility of creating one badge for each of the competences in each competence group. But this would have resulted in an excessively large number of badges that we thought would be unmanageable.

Though the BDV Badges are based upon the work produced by the EDISON project, it should be noted that they can also be related to EDSA's curriculum. Each badge has several required skills (or in EDISON's terminology "learning outcomes"), which relate to one or several of EDSA's topics. In this sense, the description given by EDSA for each topic constitutes one possible learning resources source. In fact, EDSA based their four learning pathways (data analytics, data science engineering, data management and business process management) upon four of the five competence groups in EDISON's framework [115].

6.3.4.2 Refining and Evaluating this Initial Proposal

With the aim of verifying the comprehensibility and utility of this proposal, we conducted an evaluation process which involved both industry and academia. In order to get detailed feedback and make this assessment process effective, in this initial stage we focused only on the first badge, the Data Science Analytics Badge.

Twelve companies were contacted to participate in different stages of the assessment. The aim was to obtain information about the relevance of the different required skills to their hiring practices and to ensure that the descriptions of the required skills were easy to understand.

Additionally, fifteen universities were also contacted to participate in several rounds of the evaluation. The aim was to get feedback about the review process (specifically the kinds of material to be requested of badge applicants) and about the requirements of the badge.

Fig. 6.9 Data Science Analytics Badges with academic and professional levels (v1.0)

Finally, the members of the BDVA Skills and Education Task Force were also requested to provide their opinions on the initial version of the badges as well as on the comments gathered from industry and academy throughout the entire process.

As a result of the assessment process, several changes were proposed in the initial version of the requirements of the Data Science Analytics Badge. One of the most relevant ones is the replacement of the three levels of proficiency (basic, intermediate and expert) with three different sets of required skills with two levels (academic and professional) having the same required skills. The academic level requires knowledge and training which can be acquired in an academic context, while the professional level requires real professional practice.

Furthermore, based upon comments received the descriptions of some of the requirements were modified. Some of these changes were, for example, to highlight the role of descriptive models and to separate requirements related to data preparation, data visualisation and data analytics tasks. The resulting requirements for the BDV Data Science Analytics Badge are shown below and the images of both the academic and professional badges are shown in Fig. 6.9.

Data Science Analytics Badge v1-0 Required skills

- DSA.1. Identify existing requirements to choose and execute the most appropriate data discovery techniques to solve a problem depending on the nature of the data and the goals to be achieved.
- DSA.2. Select the most appropriate techniques to understand and prepare data prior to modelling to deliver insights
- DSA.3. Assess, adapt and combine data sources to improve analytics
- DSA.4. Use the most appropriate metrics to evaluate and validate results, proposing new metrics for new applications if required
- DSA.5. Design and evaluate analysis tools to discover new relations in order to improve decision-making
- DSA.6. Use visualisation techniques to improve the presentation of the results of a data science project in any of its phases

A pilot of the entire application process to issue the Academic Level of the Data Science Analytics Badge was conducted. The goal of the pilot was to check the workflow to be followed by universities applying to issue the badge and reviewers of the applications to issue badges. Three applications were received and reviewed as part of the pilot. No major problems in the process were found and the pilot was considered to be a success. Two of the applications received as part of the pilot were accepted:

- Master's in Big Data Analytics, Universitat Politècnica de València, València, Spain
- Data and Web Science M.Sc. Programme, Aristotle University of Thessaloniki, Thessaloniki, Greece

6.3.5   Conclusions

After a survey of current trends in education in the EU was conducted, a collection of desirable properties for a data science skills recognition system was established. Those properties were compared to the benefits of existing tools for recognising skills and a hybrid between certifications and badges was selected as the appropriate tool for recognising skills in data science. Lastly, based upon the needs of stakeholders in the data science ecosystem, the details of a recognition system for data science skills were defined and the system was successfully piloted. At the moment of writing, the first open call for application to issue the Academic Level of the BDV Data Science Analytics Badge is open and can be found online at [116].

Portions of this document were taken from D4.6: A Framework for the Recognition of Data Science Skills in Europe and D4.2: Skills, Education, and Centers of Excellence Period I Report which were produced as part of the Big Data Value Ecosystem project.

This work was partially funded by the project ID: 732630 funded under: H2020-EU.2.1.1.—INDUSTRIAL LEADERSHIP—Leadership in enabling and industrial technologies - Information and Communication Technologies.

6.4   Data Management and Data Stewardship for Industry, Research and Academia

Yuri Demchenko and Steve Brewer

Establishing effective data management and data governance (DMG) in organisation is considered as a first step in digital transformation. Best practices in DMG are well defined by the DAMA (Data Management Association) and published as Data Management Body of Knowledge (DMBOK) [30], which defines a set of knowledge, competences and responsibilities of the main organisational roles and actors in the data management and governance. The DNV GL Data Quality Assessment Framework [117] and CMMI Data Management Maturity Model (DMM) [77] provide a set of industry best practice recommendations on how to achieve best use of company's data resource, converting them into assets and bringing competitive benefits. Widely adopted in research community, the FAIR data principles [78, 118] and data stewardship competence framework [57] provide a good contribution to building practically oriented DMG curricula. Below are outlines of the two distinctive courses that are related to DMG: enterprise data management and governance and research data management and stewardship.

6.4.1 Data Management and Governance. Enterprise Scope

The DMG course uses DMBOK [30] as general framework covering the majority of topics, extending them with the data science and big data analytics platforms and enriching with the FAIR and industry best practices. The following are the main topics that can be included in the course:

- Introduction. Big data infrastructure and data management and governance
- Data management concepts. Data management frameworks: DAMA Data Management Framework, the Amsterdam Information Model. Extensions for big data and data science
- Enterprise data architecture. Data lifecycle management and service delivery model. Data management and data governance activities and roles
- Data Science Professional profiles and organisational roles, skills management and capacity building
- Data architecture, data modelling and design. Data types and data models. Data modelling. Metadata. SQL and NoSQL databases overview. Distributed systems: CAP theorem, ACID and BASE
- Enterprise big data infrastructure and integration with enterprise IT infrastructure. Data warehouses. Distributed file systems and data storage
- Big data storage and platforms. Cloud-based data storage services: data object storage, data blob storage, data lakes (services by AWS, Azure, GCP)
- Trusted storage, blockchain-enabled data provenance
- FAIR data principle and data stewardship, data quality assessment and maturity model. Data repositories, open data services
- Maturity: DNV-GL Data Quality Framework, DCC RISE, CIMM, etc.
- Big data security and compliance. Data security and data protection. Security of outsourced data storage. Cloud security and compliance standards and cloud provider services assessment

The outlined topics above can be included in the practical courses for different target groups and at the different competence levels from data literacy courses to professional training and academic curricula.

6.4.2 *Research Data Management and Stewardship (RDMS)*

The research data management has numerous implementations and well supported with training materials, but in most cases, this is focused on the specific scientific domain. The courses also include growing popular FAIR principles and data stewardship-related topics [57].

The following RDMS course example is structured along practical aspects of the research data management:

A. Use cases for data management and stewardship

  (a) Preserving the scientific record

B. Data management elements (organisational and individual)

  (a) Goals and motivation for managing your data
  (b) Data formats, metadata, related standards
  (c) Creating documentation and metadata, metadata for discovery
  (d) Using data portals and metadata registries
  (e) Tracking data usage, data provenance, linked data
  (f) Handling sensitive data
  (g) Backing up data, backup tools and services
  (h) Data management plan (DMP)

C. Responsible data use (citation, copyright, data restrictions)

  (a) Data privacy and GDPR compliance

D. FAIR principles in research data management, supporting tools, maturity model and compliance

E. Data stewardship and organisational data management

  (a) Responsibilities and competences
  (b) DMP management and data quality assurance

F. Open science and open data (definition, standards, open data use and reuse, open government data)

  (a) Research data and open access
  (b) Repository and self-archiving services
  (c) RDA products and recommendations: PID, data types, data type registries, others
  (d) ORCID identifier for data and authors
  (e) Stakeholders and roles: engineer, librarian, researcher
  (f) Open data services: ORCID.org, Altmetric Doughnut, Zenodo

G. Hands-on practice includes the following topics:

  (a) Data management plan design
  (b) Metadata and tools

(c) Selection of licences for open data and contents (e.g. Creative Commons and Open Database)

6.4.3   FAIR Data Principles in Data Management

FAIR data principles (data must be findable, accessible, interoperable, reusable) [58] are growingly adopted by the research community and recognised by industry. The role of data steward is becoming important in organisational data management, which should ensure that the organisation correctly responds to tasks and challenges in managing data assets, ensuring data quality that matches organisational data-driven and data-dependent processes.

EU-funded FAIRsFAIR project is targeted to develop the FAIR data competence framework complementary to or as extension to existing and adopted data science and other competence frameworks (e.g. the ESCO-compliant EDISON Data Science Framework). The project activities are primarily focused on data science but also address other disciplines requiring data quality management. The project will develop recommendation for inclusion of the FAIR data management aspects into university curricula and professional training.

Annex: Data Science-Related Process Models

A.1 Scientific Methods and Data-Driven Research Cycle

For a data scientist who is dealing with handling data obtained in the research investigation, understanding of the scientific methods and the data-driven research cycle is essential part of knowledge that motivates necessary competences and skills for the data scientists to successfully perform their tasks and support or lead data-driven research.

The scientific method is a body of techniques for investigating phenomena, acquiring new knowledge or correcting and integrating previous knowledge [119, 120]. Traditional steps of the scientific research were developed over time since the time of ancient Greek philosophers through modern theoretical and experimental research where experimental data or simulation results were used to validate the hypothesis formulated based on initial observation or domain knowledge study. The general research methods include observational methods, opinion-based methods and experimental and simulation methods.

The increased power of computational facilities and advent of big data technologies created a new paradigm of the data-driven research that enforced the ability of researchers to make observation of the research phenomena based on bigger datasets and applying data analytics methods to discover hidden relations and processes not available to deterministic human thinking. The principles of the data-driven research were formulated in the seminal work *The Fourth Paradigm: Data-Intensive Scientific Discovery* edited by Tony Hey [121].

The research process is iterative by its nature and allows scientific model improvement by using continuous research cycle that typically includes the following eight basic stages:

1. Define research questions
2. Design experiment representing initial model of research object or phenomena
3. Collect data
4. Analyse data

J. J. Cuadrado-Gallego, Y. Demchenko (eds.), *The Data Science Framework*,
https://doi.org/10.1007/978-3-030-51023-7

5. Identify patterns
6. Hypothesise explanation
7. Test hypothesis
8. Refine model and start new experiment cycle

The traditional research process may be concluded with the scientific publication and archiving of collected data. Data-driven and data-powered/driven research paradigm allows research data reuse and combining them with other linked datasets to reveal new relations between initially not linked processes and phenomena. As an example, biodiversity research when studying specific species population can include additional data from weather and climate observation, solar activity, another species migration and technogenic factor. The proposed CF-DS introduces research methods as an important component of the data science competences and knowledge and uses data life cycle as an approach to define the data management-related competences group.

A.2   Scientific Data Lifecycle Management Model

Data life cycle is an important component of data-centric applications, which data science and big data applications belong to. Data lifecycle analysis and definition is addressed in many domain-specific projects and studies. Extensive compilation of the data lifecycle models and concepts is provided in the CEOS.WGISS.DSIG document [122].

For the purpose of defining the major groups of competences required for the data scientist working with scientific applications and data analysis, we will use the Scientific Data Lifecycle Management (SDLM) model [123] shown in Fig. A.1a defined as a result of analysis of the existing practices in different scientific communities. Figure A.1b illustrates the more general Big Data Lifecycle Management (BDLM) model involving the main components of the Big Data Reference Architecture defined in NIST BDIF [3, 124, 125]. The proposed models are sufficiently generic and compliant with the data lifecycle study results presented in [122].

The generic scientific data life cycle includes a number of consequent stages: research project or experiment planning; data collection; data processing; publishing research results; discussion, feedback; archiving (or discarding). SDLM reflects complex and iterative process of the scientific research that is also present in data science analytics applications.

Both SDLM and BDLM require data storage and preservation at all stages which should allow data reuse/re-purposing and secondary research on the processed data and published results. However, this is possible only if the full data identification, cross-reference and linkage are implemented in scientific data infrastructure (SDI). Data integrity, access control and accountability must be supported during the whole data life cycle.

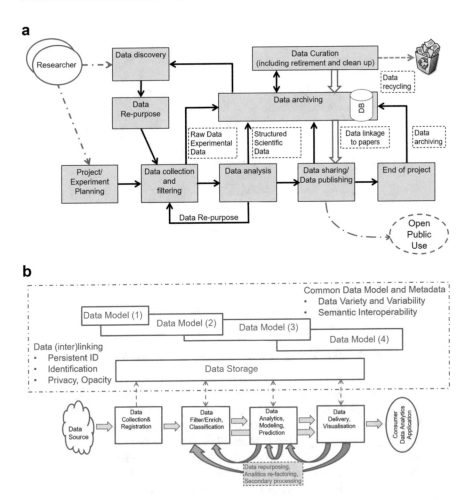

Fig. A.1 Data Lifecycle Management in (**a**) Scientific data lifecycle management—e-Science focused, (**b**) Big Data Lifecycle Management model (compatible with NIST NBDIF)

Data curation is an important component of the discussed data lifecycle models and must also be done in a secure and trustworthy way. The research data management and handling issues are extensively addressed in the work of the Research Data Alliance.

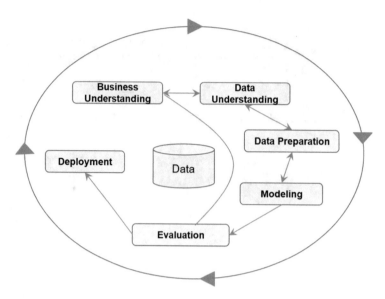

Fig. A.2 CRISP-DM (Cross-Industry Standard Process for Data Mining)

A.3 Cross-Industry Standard Process for Data Mining

Although initially proposed in 1990s, CRISP-DM (Cross-Industry Standard Process for Data Mining) [126] model is still used in defining data mining and data analytics workflows and processes. It is also used for defining common data mining and data analytics processes and stages, but not limited to data analytics or data management. Figure A.2 illustrates CRISP-DM stages. It is important to mention that modern agile technologies and agile business technologies engage the main data handling and data analytics processes into continuous development and continuous improvement cycle.

Table A.1 provides example of initial mapping CF-DS competences to the CRISP-DM processes and stages that will need to undergo cross-checking with the corresponding knowledge subjects in DS-BoK

A.4 Business Process Management Lifecycle

New-generation agile data-driven enterprises (ADDE) use data science methods to continuously monitor and improve their business processes and services. The data-driven business management model allows combining different data sources to improve predictive business analytics which allows making more effective solutions, faster adaptation of services and more specifically targeting different customer

Table A.1 Mapping CF-DS competences to CRISP-DM process and stages

CRISP-DM processes and stages	Description	Mapping to CF-DS
Business understanding	General business understanding, role of data and required actionable information	DSBAxx DSRMPxx
Determine business objectives	Business objectives (SMART approach). Specific, measurable, attainable (in principle), relevant and timely. This is performed by business stakeholders!	DSBA01
	Business success criteria (or benchmark or threshold values)	
Assess situation	Inventory of resources, requirements, assumptions and constraints	DSBA01
	Risks and contingencies	
	Costs and benefits	
Determine data mining goals	Data mining goals	DSRMP05, DSRMP06
	Data mining success criteria	
Produce project plan	Project plan	DSRMP05, DSRMP06
	Initial assessment of tools and techniques	
Data acquisition and understanding	Collect data, assign metadata, explore data, run ETL processes	DSDAxx DSDMxx
Collect initial data	Acquire access to data from internal and external sources (API, webscrapping). In a steady state, data extraction and transfer routines would be in place	DSDA03
Describe data	Describe data, add metadata	DSDA03 DSDM04
Explore data	Checking on definitions and meaning of data acquired. This requires business knowledge (business analyst/or business stakeholder)	DSDA03 DSBA01
	Examine the "surface" properties of the acquired data	
	Understand distribution of key attributes, perform initial visualisation, understand initial relationships between small number of attributes, perform simple aggregations	
Verify data quality	Checking if data is up-to-date	DSDA03
	Checking if data is complete, correct, error-free	
Data preparation (select and cleanse)	Data preprocessing, cleaning, reduction, sampling	DSDAxx
Select data	Decide on data to be used for analysis	DSDA03
Clean data	Increase data quality by substitution, imputation (estimating of missing data)/insertion of suitable defaults Identification of outliers, anomalies and patterns	DSDA03, DSDM05
Construct data	Transform dataset, produce derived values, produce new (composed) records	DSDA03
Integrate data	Merging of tables (joins) or aggregations of data	DSDA03, DSDM03

(continued)

Table A.1 (continued)

CRISP-DM processes and stages	Description	Mapping to CF-DS
Format data	Syntactic modifications (not changing meaning but produce format required by modelling tool, e.g. convert dataset to JSON)	DSDA03, DSDM03
Hypothesis and modelling		DSDAxx
Select modelling techniques	Decide on techniques to be used, depending on type of problem (ML, decision tree, neural nets, etc.)	DSDA01
Generate test design	Generate procedure or mechanism to test model quality and validity Separate dataset in train and validation sets and test sets	DSDA01
Build model	Run the modelling tool on the prepared dataset to create one or more models. Perform parameter selection (e.g. hyperparameter)	DSDA01, DSDA02
Assess model and revise parameters	Judge success of the application of modelling and discovery technically: contact business analysts and domain experts in order to discuss model. Summarise qualities of generated models (i.e. accuracy).	DSDA04
	Revise parameter settings and tune them for the next run in the build model task	
Evaluate results	Summarise assessment results in terms of business success criteria This involves business stakeholders. This is not to evaluate the model accuracy/generalisation: this is already done in the previous step	DSDA04
Review process and determine improvement	Formally assess data analytics process	DSDA04, DSRM01
Deployment, operations and maintenance	Deploy application or process, maintain, prepare	DSRM06
Plan deployment	Determine strategy for deployment, determine how information will be propagated to users, decide how the use of result will be monitored and benefits measured	DSRM06
	Plan potential re-coding (e.g. from python to Java for production environment)	
Plan monitoring and maintenance	Check for dynamic aspects, decide how accuracy will be monitored, determine threshold below, the result of which cannot be used anymore or should be updated/recalibrated Monitor and measure performance of model	DSDA05, DSRMP06
	Plan for DEVOPS or continuous delivery/agile development	DSENGxx
Produce final report	Produce final report, create dashboard for proper visualisation	DSDA06, DSRM06

groups as well as doing optimal resources allocations depending on market demand and customer incentives.

Similarly, to the research domain the data-driven methods and technologies change how the modern business operates attempting to benefit from the new insight that big data can give into business improvement including internal processes organisation and relation with customers and market processes. Understanding business process management life cycle [127] is important to identify necessary competences and knowledge for business-oriented data science profiles.

The following are typical six stages of the business process management life cycle: (1) Define the business target: both services and customers; (2) design the business process; (3) model/plan; (4) deploy and execute; (5) monitor and control; and (6) optimise and redesign.

The need for the business process management competences and knowledge for business-oriented data science profiles is in CF-DS.

References

1. D. Donoho, "50 years of Data Science," 2015. [Online]. Available: https://www.tandfonline.com/doi/full/10.1080/10618600.2017.1384734.
2. National Institute of Standards and Technologies, NIST, [Online]. Available: https://www.nist.gov/.
3. "NIST SP 1500-1 NIST Big Data interoperability Framework (NBDIF): Volume 1: Definitions, September 2015 [online]," National Institute of Standards and Technologies, NIST, [Online]. Available: http://nvlpubs.nist.gov/nistpubs/SpecialPublications/NIST.SP.1500-1.pdf.
4. Research Data Alliance, RDA, [Online]. Available: https://www.rd-alliance.org/.
5. RDA Interest Group on Education and Training on Handling Research Data, IG-ETHRD, [Online]. Available: https://www.rd-alliance.org/ig-education-and-training-handling-research-data-rda-9th-plenary-meeting.
6. EDISON Data Science Framework Community Initiative, [Online]. Available: https://github.com/EDISONcommunity/EDSF/wiki/EDSFhome.
7. "ESCO handbook: European Skills, Competences, Qualifications and Occupation," European Commission, 2017. [Online]. Available: https://ec.europa.eu/esco/portal/document/en/0a89839c-098d-4e34-846c-54cbd5684d24.
8. "European e-Competence Framework," European Commission, [Online]. Available: http://www.ecompetences.eu.
9. "European e-Competence Framework 3.0. A common European Framework for ICT Professionals in all industry sectors. CWA 16234:2014 Part 1," 2014. [Online]. Available: http://ecompetences.eu/wp-content/uploads/2014/02/European-e-Competence-Framework-3.0_CEN_CWA_16234-1_2014.pdf.
10. "User guide for the application of the European e-Competence Framework 3.0. CWA 16234:2014 Part 2." [Online]. Available: http://ecompetences.eu/wp-content/uploads/2014/02/European-e-Competence-Framework-3.0_CEN_CWA_16234-2_2014.pdf.
11. "European ICT Professional Profiles CWA 16458 (2012) (Updated by e-CF3.0)," 2012. [Online]. Available: http://relaunch.ecompetences.eu/wp-content/uploads/2013/12/EU_ICT_Professional_Profiles_CWA_updated_by_e_CF_3.0.pdf.
12. "European Qualifications Framework (EQF)," [Online]. Available: https://ec.europa.eu/ploteus/content/descriptors-page.
13. "Information Technology Competency Model of Core Learning Outcomes and Assessment for Associate-Degree Curriculum," 2014. [Online]. Available: http://www.capspace.org/uploads/ACMITCompetencyModel14October2014.pdf.
14. "The 2012 ACM Computing Classification System." [Online]. Available: http://www.acm.org/about/class/class/2012.

J. J. Cuadrado-Gallego, Y. Demchenko (eds.), *The Data Science Framework*,
https://doi.org/10.1007/978-3-030-51023-7

15. "Bloom's taxonomy: the 21st century version." [Online]. Available: http://www. educatorstechnology.com/2011/09/blooms-taxonomy-21stcentury-version.html.
16. "The United States Department of Labor IT Competency Model," [Online]. Available: www. careeronestop.org/COMPETENCYMODEL/pyramid.aspx?IT=Y.
17. M. V. Harris, "Analysing the Analysers, O'Reilly Strata Survey," 2013. [Online]. Available: http://cdn.oreillystatic.com/oreilly/radarreport/0636920029014/Analyzing_the_Analyzers. pdf.
18. "Big Data Analytics: Assessment of demand for Labour and Skills 2013-2020." Tech Partnership publication, SAS UK & Ireland, 2014. [Online]. Available: https://www.e-skills.com/ Documents/Research/General/BigData_report_Nov14.pdf.
19. DARE Project Recommended Data Science and Analytics Skills, Working report, 2017.
20. "LERU Roadmap for Research Data," LERU Research Data Working Group, 2013. [Online]. Available: http://www.leru.org/files/publications/AP14_LERU_Roadmap_for_Research_ data_final.pdf.
21. PwC and BHE, "PwC and BHEF report "Investing in America's data science and analytics talent: The case for action"," PwC and BHE, 2017. [Online]. Available: http://www.bhef.com/ publications/investing-americas-data-science-and-analytics-talen.
22. "Burning Glass Technology, IBM, and BHEF report "The Quant Crunch: How the demand for Data Science Skills is disrupting the job Market"," 2017. [Online]. Available: http://www. bhef.com/publications/quant-crunch-how-demand-data-science-skills-disrupting-job-market.
23. Meeting of the OECD Council at Ministerial Level, "Going Digital in a Multilateral World, Paris OECD Report, 2018," 2018. [Online]. Available: https://www.oecd.org/going-digital/C-MIN-2018-6-EN.pdf.
24. "P21's Framework for 21st Century Learning," [Online]. Available: http://www.p21.org/ storage/documents/P21_framework_0515.pdf.
25. "ACM and IEEE Computer Science Curricula 2013 (CS2013) [online]," [Online]. Available: https://doi.org/10.1145/2534860.
26. "ACM Curricula recommendations." [Online]. Available: http://www.acm.org/education/cur ricula-recommendations.
27. "ICT professional Body of Knowledge (ICT-BoK)." [Online]. Available: http://www.ictbok. eu/images/EU_Foundationa_ICTBOK_final.pdf.
28. "Software Engineering Body of Knowledge (SWEBOK)." [Online]. Available: https://www. computer.org/web/swebok/v3.
29. "Business Analytics Body of Knowledge (BABOK)." [Online]. Available: http://www.iiba. org/babok-guide.aspx.
30. "Data Management Body of Knowledge (DM-BoK) by Data Management Association International (DAMAI)." [Online]. Available: http://www.dama.org/sites/default/files/download/ DAMA-DMBOK2-Framework-V2-20140317-FINAL.pdf.
31. "Project Management Professional Body of Knowledge (PM-BoK)." 2017. [Online]. Available: http://www.pmi.org/PMBOK-Guide-and-Standards/pmbok-guide.aspx.
32. ISO/IEC, "Software Engineering — Guide to the software engineering body of knowledge. TR 19759:2015 [ISO/IEC TR 19759:2015]," [Online]. Available: https://www.iso.org/stan dard/67604.html.
33. "ISO 21500:2012 Guidance on Project Management," 2012. [Online]. Available: https:// www.iso.org/standard/50003.html.
34. "Information Technology Competency Model of Core Learning Outcomes and Assessment for Associate-Degree Curriculum (2014)," [Online]. Available: http://www.capspace.org/ uploads/ACMITCompetencyModel14October2014.pdf.
35. "European Credit Transfer and Accumulation System (ECTS)." [Online]. Available: http://ec. europa.eu/education/ects/users-guide/docs/year-2009/ects-users-guide-2009_en.pdf.
36. "Carnegie unit credit hour." [Online]. Available: https://www.luminafoundation.org/files/ resources/carnegie-unit-report.pdf.

37. "Information Technology Competency Model of Core Learning Outcomes and Assessment for Associate-Degree Curriculum," 2014. [Online]. Available: http://capspace.org/uploads/ACMITCompetencyModel14October2014.pdf.

38. "European Guidelines for Curriculum Development for e-Leadership Skills: Proposal for further development and adoption [on-line]," January 2016. [Online]. Available: http://eskills-scale.eu/fileadmin/eskills_scale/all_final_deliverables/scale_d2_working_document_cp_guidelines.pdf.

39. B. S. Bloom, M. D. Engelhart, E. J. Furst, W. H. Hill and D. R. Krathwohl, Taxonomy of educational objectives: The classification of educational goals. Handbook I: Cognitive domain, New York: David McKay Company, 1956.

40. L. Anderson and D. Krathwohl, A taxonomy for learning, teaching, and assessing, Abridged Edition. Boston, MA: Allyn and Bacon, 2001.

41. T. W. Wlodarczyk and T. J. Hacker, "Problem-Based Learning Approach to a Course in Data Intensive Systems," in *Cloud Computing Technology and Science (CloudCom), IEEE 6th International Conference on. IEEE, 2014*, 2014.

42. D. A. Kolb, Experiential learning: experience as the source of learning and development. Prentice-Hall, 1984.

43. P. Blumenfeld, E. Soloway, R. Marx, J. Krajcik, M. Guzdial and A. Palincsar, "Motivating Project-Based Learning: Sustaining the Doing, Supporting the Learning," *Educational Psychologist, vol. 26, no. 3–4, pp. 1991* vol. 26, no. 3–4, p. 369–398, 1991.

44. T. Malone and M. Lepper, "Making learning fun: A taxonomy of intrinsic motivations for learning," *Aptitude, learning, and instruction,* vol. 3, p. 223–253, 1987.

45. J. Biggs, "Enhancing teaching through constructive alignment," *Higher education,* vol. 32, p. 347–364, 1996.

46. M. Ben-Ari, "Constructivism in computer science education," *Journal of Computers in Mathematics and Science Teaching,* vol. 20, pp. 45–73, 2001.

47. "The Aalborg Model for Problem Based Learning (PBL)," [Online]. Available: http://www.en.aau.dk/education/problem-based-learning/.

48. A. S. Thackaberry, "A CBE Overview: The Recent History of CBE." [Online]. Available: http://evolllution.com/programming/applied-and-experiential-learning/a-cbe-overview-the-recent-history-of-cbe.

49. "International Standard Classification of Occupations (ISCO) Home," [Online]. Available: http://www.ilo.org/public/english/bureau/stat/isco/.

50. "ISCO-88, the International Standard Classification of Occupations," [Online]. Available: http://www.ilo.org/public/english/bureau/stat/isco/isco88/index.htm.

51. "What is a data scientist? 14 definitions of a data scientist!," [Online]. Available: http://bigdata-madesimple.com/what-is-a-data-scientist-14-definitions-of-a-data-scientist/.

52. "LinkedIn's Daniel Tunkelang On "What Is a Data Scientist?"," [Online]. Available: http://www.forbes.com/sites/danwoods/2011/10/24/linkedins-daniel-tunkelang-on-what-is-a-data-scientist/.

53. "Realising the European Open Science Cloud. First report and recommendations of the Commission High Level Expert Group on the European Open Science Cloud, October 2016," [Online]. Available: https://ec.europa.eu/research/openscience/pdf/realising_the_european_open_science_cloud_2016.pdf.

54. "GO FAIR Initiative." [Online]. Available: https://www.dtls.nl/fair-data/go-fair/.

55. "FAIR Data principles." [Online]. Available: https://www.dtls.nl/fair-data/fair-principles-explained/.

56. "Stewardship Skills for Open Science and Scholarship, EOSCpilot Workpackage WP7." [Online]. Available: https://www.eoscpilot.eu/themes/wp7-skills.

57. "EOSCpilot Deliverable D7.3: Stewardship Skills for Open Science and Scholarship, EOSCpilot Project deliverable, June 2018 [online]," [Online]. Available: https://eoscpilot.eu/sites/default/files/eoscpilot-d7.3.pdf.

58. "FIARsFAIR Project: Fostering FAIR data principles in Europe." [Online]. Available: https://www.fairsfair.eu/.
59. "Cortnie Abercrombie, What CEOs want from CDOs and how to deliver on it." [Online]. Available: http://www.slideshare.net/IBMBDA/what-ceos-want-from-cdos-and-how-to-deliver-on-it.
60. "PwC and BHEF report "Investing in America's data science and analytics talent: The case for action"," 2017. [Online]. Available: http://www.bhef.com/publications/investing-americas-data-science-and-analytics-talent .
61. "Burning Glass Technology, IBM, and BHEF report "The Quant Crunch: How the demand for Data Science Skills is disrupting the job Market"," 2017. [Online]. Available: http://www.bhef.com/publications/quant-crunch-how-demand-data-science-skills-disrupting-job-market .
62. Y. Demchenko, L. Comminiello and G. Reali, "Designing Customizable Data Science Curriculum using Ontology for Science and Body of Knowledge," in *International Conference on Big Data and Education (ICBDE2019)*, London, United Kingdom, March 30–April 1, 2019.
63. Y. Demchenko, A. Belloum, C. de Laat, C. Loomis, T. Wiktorski and E. Spekschoor, "Customizable Data Science Educational Environment: From Competences Management and Curriculum Design to Virtual Labs On-Demand," in *Proc. of the IEEE International Conference and Workshops on Cloud Computing Technology and Science (CloudCom2017)*, Hong Kong, 11–14 December 2017.
64. T. Wiktorski, Y. Demchenko and O. Chertov, "Curriculum Implementation for Various Types of Big Data Infrastructure Courses," in *Proc. 5th IEEE STC CC Workshop on Curricula and Teaching Methods in Cloud Computing, Big Data and Data Science (DTW2019)*, San Diego, California, USA, September 24–27, 2019.
65. Y. Demchenko, "Big Data Platforms and Tools for Data Analytics in the Data Science Engineering Curriculum," in *Proc of the 3rd International Conference on Cloud and Big Data (ICCBDC 2019)*, Oxford, UK, August 28–30, 2019.
66. Y. Demchenko, T. Wiktorski, S. Brewer and J. J. Cuadrado-Gallego, "EDISON Data Science Framework (EDSF) Extension to Address Transversal Skills required by Emerging Industry 4.0 Transformation," in *Proc. of the 5th IEEE STC CC Workshop on Curricula and Teaching Methods in Cloud Computing, Big Data and Data Science (DTW2019), part of the eScience 2019 Conference*, San Diego, California, USA, September 24–27, 2019.
67. Y. Demchenko and S. Brewer , "Data Management and Governance Courses in Academic Curricula," in *CODATA-Helsinki Workshop on FAIR RDM in institutions 2019*, Helsinki, Finland, 20–21 October 2019.
68. European Commission, EDC-IC, "Final results of the European Data Market study measuring the size and trends of the EU data economy," March 2017. [Online]. Available: https://ec.europa.eu/digital-single-market/en/news/final-results-european-data-market-study-measuring-size-and-trends-eu-data-economy.
69. OECD, "OECD Skills Outlook 2019," May 2019.
70. M. Horridge, S. Jupp, G. Moulton, A. Rector, R. Stevens and C. Wroe, A Practical Guide to Building OWL Ontologies using Protege 4 and CO-ODE Tools.
71. J. Han Lan and T. Baldwin, "An Empirical Evaluation of doc2vec with Practical Insights into Document Embedding Generation," in *Proceedings of the 1st Workshop on Representation Learning for NLP*, Berlin, Germany, 2016.
72. Q. Le and T. Mikolov, "Proceedings of the 31st International Conference on Machine Learning, PMLR 32(2)," Beijing, China, June 2014.
73. P. Lord, "Ontogenesis. Components of an Ontology." 2010. [Online]. Available: http://ontogenesis.knowledgeblog.org/514/.
74. [Online]. Available: https://www.kagle.com.
75. [Online]. Available: https://archive.ics.uci.edu/ml/index.php.
76. Y. Demchenko and et al. "EDISON Data Science Framework: A Foundation for Building Data Science Profession for Research and Industry," in *Proc. of the 8th IEEE International*

Conference and Workshops on Cloud Computing Technology and Science (CloudCom2016), Luxemburg, 12–15 December 2016.
77. CMMI Institute, "Data Maturity Model," 2018.
78. B. Mons and et al. "The FAIR guiding principles for scientific data management and stewardship," [Online]. Available: https://www.nature.com/articles/sdata201618.
79. "Apache Hadoop," [Online]. Available: https://hadoop.apache.org/.
80. "Hadoop Ecosystem and their components - A Complete Tutorial," [Online]. Available: https://data-flair.training/blogs/hadoop-ecosystem-components/.
81. "Apache Hive Tutorial," [Online]. Available: https://cwiki.apache.ord/confluence/display/Hive/Tutorial.
82. "Apache Pig Tutorial," [Online]. Available: https://data-flair.training/blogs/hadoop-pig-tutorial/.
83. "Amazon Web Services (AWS)," [Online]. Available: https://aws.amazon.com.
84. "Microsoft Azure," [Online]. Available: https://docs.microsoft.com/en-us/azure/architecture/data-guide/.
85. "Google Cloud Platform," [Online]. Available: https://cloud.google.com.
86. "Cloudera Hadoop Cluster (CDH)," [Online]. Available: https://www.cloudera.com/documentation/other/reference-architecture.html.
87. "Hortonworks Data Platform," [Online]. Available: https://hortonworks.com/products/data-platforms/hdp/.
88. Y. Demchenko, D. Bernstein, A. Oprescu, T. W. Wlodarczyk and C. de Laat, "New Instructional Models for Building Effective Curricula on Cloud Computing Technologies and Engineering," in *Proc. of the 5th IEEE International Conference and Workshops on Cloud Computing Technology and Science (CloudCom2013)*, Bristol, UK, 2–5 December 2013.
89. [Online]. Available: https://eur-lex.europa.eu/legal-content/EN/TXT/PDF/?uri=CELEX:32018H1210(01)&from=EN.
90. [Online]. Available: http://europa.eu/youreurope/citizens/education/university/recognition/index_en.htm.
91. [Online]. Available: http://europa.eu/youreurope/citizens/work/professional-qualifications/recognition-of-professional-qualifications/index_en.htm .
92. [Online]. Available: http://eur-lex.europa.eu/legal-content/EN/TXT/PDF/?uri=CELEX:52009XG0528(01)&from=EN.
93. [Online]. Available: http://eur-lex.europa.eu/legal-content/EN/TXT/PDF/?uri=CELEX:52015XG1215(02)&from=EN.
94. [Online]. Available: Cedefop (2015). European guidelines for validating non-formal and informal learning. Luxembourg: Publications Office. Cedefop reference series; No 104. https://doi.org/10.2801/008370.
95. [Online]. Available: https://ec.europa.eu/education/resources-and-tools/european-credit-transfer-and-accumulation-system-ects_en.
96. [Online]. Available: https://ec.europa.eu/education/resources-and-tools/the-european-credit-system-for-vocational-education-and-training-ecvet_en.
97. [Online]. Available: http://ec.europa.eu/social/main.jsp?catId=1223.
98. [Online]. Available: https://eur-lex.europa.eu/legal-content/EN/TXT/PDF/?uri=CELEX:32004D2241&from=EN .
99. [Online]. Available: https://eur-lex.europa.eu/legal-content/EN/TXT/PDF/?uri=CELEX:52013DC0899&from=en.
100. [Online]. Available: https://ec.europa.eu/transparency/regdoc/rep/1/2016/EN/1-2016-625-EN-F1-1.PDF.
101. [Online]. Available: http://eurlex.europa.eu/LexUriServ/LexUriServ.do?uri=OJ:C:2008:111:0001:0007:EN:PDF.
102. [Online]. Available: https://ec.europa.eu/esco/portal/document/mt/89a2ca9a-bc79-4b95-a33b-cf36ae1ac6db.
103. [Online]. Available: https://www.iso.org/certification.html.

104. [Online]. Available: https://en.wikipedia.org/wiki/Academic_degree.
105. [Online]. Available: https://en.wikipedia.org/wiki/List_of_unaccredited_institutions_of_higher_education.
106. [Online]. Available: https://en.wikipedia.org/wiki/Academic_certificate.
107. [Online]. Available: https://www.extension.harvard.edu/academics/professional-graduate-certificates/data-science-certificate .
108. [Online]. Available: http://datascience.uci.edu/data-science-certificate-program/ .
109. [Online]. Available: https://scs.georgetown.edu/programs/375/certificate-in-data-science/ .
110. [Online]. Available: http://extension.berkeley.edu/public/category/courseCategoryCertificateProfile.do?method=load&certificateId=28652248.
111. "Open digital badges." *The Electronic Journal of English as a Second Language,* vol. 18, no. 1, pp. 1–11.
112. [Online]. Available: https://www.ibm.com/services/learning/M425350C34234U21.
113. [Online]. Available: https://www.purdue.edu/cie/globallearning/badges.html.
114. [Online]. Available: http://stackoverflow.com/help/badges/.
115. [Online]. Available: http://courses.edsa-project.eu/course/view.php?id=70 (Consulted January 2020).
116. [Online]. Available: https://www.big-data-value.eu/skills/skills-recognition-program/call-for-academic-level-data-science-analytics-badge-issuers.
117. "DNVGL-RP-0497 Data quality assessment framework," 2017. [Online]. Available: https://rules.dnvgl.com/docs/pdf/DNVGL/RP/2017-01/DNVGL-RP-0497.pdf.
118. "Turning FAIR into reality. Final Report and Action Plan from the European Commission Expert Group on FAIR Data," 2018. [Online]. Available: https://ec.europa.eu/info/sites/info/files/turning_fair_into_reality_1.pdf.
119. W. E. Bright Jr., An Introduction to Scientific Research, Dover Publications, 1991.
120. "Research Methodology." [Online]. Available: https://explorable.com/research-methodology.
121. T. Hey, S. Tansley and K. Tolle, Eds., The Fourth Paradigm: Data-Intensive Scientific Discovery, Microsoft Corporation.
122. Data Life Cycle Models and Concepts, CEOS Version 1.2. Doc. Ref.: CEOS.WGISS.DSIG, 2012.
123. "European Union. A Study on Authentication and Authorisation Platforms For Scientific Resources in Europe. Internal identification," European Commission, 2012. [Online]. Available: http://cordis.europa.eu/fp7/ict/e-infrastructure/docs/aaa-study-final-report.pdf.
124. Y. Demchenko, P. Membrey, P. Grosso and C. de Laat, "Addressing Big Data Issues in Scientific Data Infrastructure," in *First International Symposium on Big Data and Data Analytics in Collaboration (BDDAC 2013). Part of The 2013 International Conference on Collaboration Technologies and Systems (CTS 2013),* San Diego, California, USA, 2013.
125. "NIST SP 1500-6 NIST Big Data interoperability Framework (NBDIF): Volume 6: Reference Architecture," 2015. [Online]. Available: http://nvlpubs.nist.gov/nistpubs/SpecialPublications/NIST.SP.1500-6.pdf.
126. "Cross Industry Standard Process for Data Mining (CRISP-DM) Reference Model," [Online]. Available: http://crisp-dm.eu/reference-model/.
127. T. Panagacos, "The Ultimate Guide to Business Process Management: Everything you need to know and how to apply it to your organization," *CreateSpace Independent Publishing Platform,* 2012.
128. [Online]. Available: http://eskills4jobs.ec.europa.eu/.

Printed in the United States
by Baker & Taylor Publisher Services